MAGIC TRICKS, CARD SHUFFLING AND DYNAMIC COMPUTER MEMORIES

MAGIC TRICKS, CARD SHUFFLING AND DYNAMIC COMPUTER MEMORIES

S. BRENT MORRIS

Mathematical Association
merica

WITHDRAWN

Cartoons and cover illustrations by John Johnson, Teapot Graphics.
Card shuffling illustrations by Earle Oakes.
Set in type by Integre Technical Publishing Co., Inc.

ISBN 0-88385-527-5

Printed in the United States of America

Current Printing (last digit):
10 9 8 7 6 5 4 3 2 1

SPECTRUM SERIES

Published by
THE MATHEMATICAL ASSOCIATION OF AMERICA

———

SPECTRUM SERIES

The Spectrum Series of the Mathematical Association of America was so named to reflect its purpose: to publish a broad range of books including biographies, accessible expositions of old or new mathematical ideas, reprints and revisions of excellent out-of-print books, popular works, and other monographs of high interest that will appeal to a broad range of readers, including students and teachers of mathematics, mathematical amateurs, and researchers.

MAA Service Center
P. O. Box 91112
Washington, DC 20090-1112
800-331-1MAA   FAX 301-206-9789

To Mother
who always was willing to choose a card, any card.

But in shewing feats, and juggling with cards, the principall point consisteth in shuffling them nimblie.... Hereby you shall seeme to worke woonders....

Reginald Scott, 1584
The Discoverie of Witchcraft

THE WIZARD OF ID   Brant parker and Johnny hart

by permission of Johnny Hart and Creators Syndicate, Inc.

Contents

Author's Preface:
Evidence of a Misspent Youth

This book began a long time before I ever thought of shuffling a deck of cards. Some time in the mid- to late 1950s I was watching the *Howdy Doody Show*. Buffalo Bob Smith, the emcee, did a magic trick with a monkey: He lit a fire in a chafing dish, put the cover on to extinguish the flames, and, upon removing the cover, the dish was filled with popcorn. I was amazed. I was enthralled. I was hooked. My life's career was set—I was going to be a magician!

At age eight I had my professional debut: a nickel-a-head show in the garage. At twelve I moved into the more prestigious venue of entertaining at neighborhood birthday parties. At about sixteen I learned about the perfect shuffle. A magician friend told me that eight perfect shuffles returned a deck to its original order. I can still remember buying a new deck of cards and trying direct experimentation to confirm or deny this bold claim. Much to my delight and surprise, the claim was true.

About the same time another magician friend (or perhaps the same one who told me about eight shuffles) showed me how to move the top card to any position in the deck with perfect shuffles. The binary representation of the final location gives a sequence of in- and out-shuffles that moves the top card. The magician has to convert quickly from decimal to binary while performing, but a good magic trick makes learning binary arithmetic worthwhile.

When I was eighteen I was president of the Thomas Jefferson H.S. Math Club in Dallas, Texas. Dr. Patricia Copley was the club sponsor and very influential in my decision to pursue mathematics. In 1968 I presented a program on mathematics and magic and explained, among other things, the card-placement trick with binary numbers. Three years later I was Vice-President of the S.M.U. chapter of Kappa Mu Epsilon, a math honorary fraternity. I volunteered to present a program on mathematics and magic. I included more sophisticated effects than in my high school presentation, but the card placement with perfect shuffles remained. It was a good trick, and its math is nontrivial. Along the way I started reading articles on the perfect shuffle, including those by Martin Gardner and Solomon Golomb.

Two years later I was a graduate student at Duke University taking a course in number theory from Prof. Leonard Carlitz. At the end of the semester he assigned a term paper—an unusual (and disquieting) development in a math course. At Christmas break I pondered what to write on, and finally decided to write up the card-placement trick that had served me so well in high school and as an undergraduate. When he returned my paper, Dr. Carlitz said it was nice and I should try to get it published. I submitted it to the *Pi Mu Epsilon Journal*, and it was soon in print.

That spring Prof. Robert Hartwig from North Carolina State University visited Dr. Carlitz to discuss a problem. Prof Hartwig had been working for a couple of years with a class of permutation matrices, but he was unable to determine their characteristic and minimal polynomials. Dr. Carlitz showed the problem to me and suggested I look at it. That evening as I worked on the problem, it dawned on me that the permutations were generalized perfect shuffles. Instead of dividing a deck into two packets and interlacing them, the deck was divided into k packets and then interlaced. I recognized the problem as an old friend, and quickly provided a solution for Prof. Hartwig.

During that summer and the next fall Prof. Hartwig and I worked together on permutation matrices. Our efforts resulted in two publications. While this research was underway, I was completing the coursework and preliminary examination for my doctorate. With those hurdles behind me, I made an appointment with Prof. Carlitz, whom I had chosen as a dissertation advisor, to discuss my dissertation problem. Choosing a problem is a serious step in the life of any mathematician and represents the focused study of the last two to three years of graduate school.

When I met with Prof. Carlitz he asked me what kind of math I enjoyed. I quickly replied that the permutation group work with Prof. Hartwig had been fun and that I would like to pursue it some more. Dr. Carlitz's reply is still burned in my memory: "Oh, if that's what you want to do, then you're through with your Ph.D. Just write an introduction and some connecting material to go with your papers and you've finished your dissertation." I could only muster a stunned "Thank you" as I left his office that day.

Dr. Carlitz was good to his word, and I finished my dissertation that semester. I think the typing was a greater burden for my wife Nancy than the final writing was for me. In any event I now have the distinction of being the only person in the world (as far as I know) with a Ph.D. in card shuffling. After graduating from Duke, my first assignment as a professional mathematician led me back to card shuffling. Glenn Stahley, my supervisor, showed me an article from *IEEE Transactions on Computers* in which perfect shuffles had been used to design a computer memory. I instantly recognized my favorite card placement trick but in an unexpected setting. I worked on this memory problem with two colleagues, Art Valliere and Rich Wisniewski. Our research produced an article and a patent, but, alas, no vastly-improved computer chips.

This book is a leisurely study of the mathematics of card shuffling that has fascinated and served me well for so many years. (Two presentations to math clubs, a term paper, a doctoral dissertation, four refereed publications, a patent, an MAA short course, and scores of invited colloquia is good service indeed.) I have tried to write a "friendly" book for readers who majored in math ten years ago or who are willing to do a little extra study to learn about some good tricks. I will have succeeded if my magician friends complain the book is too mathematical and my mathematician friends worry there's too much performance detail.

Happy shuffling!

S. Brent Morris
Columbia, MD
December 1997

Acknowledgments

Pages make up a book just as many different cards make up a deck. But filling the pages of this book would not have been possible without the help of many resources. I gratefully acknowledge the assistance of

- Art Benjamin, for making my text read better than I deserve;
- Ben Cole, for capturing my shuffles on film;
- George Dailey, for time and again opening his library to me;
- Peter Denning, for suggesting the format for the descriptions of tricks;
- Persi Diaconis, for documenting so many early shuffle references;
- Paul Gertner, for letting me use his trick "Unshuffled";
- Earle Oakes, for clarifying the sleight of hand with his artwork;
- Jon Racherbaumer, for letting me use the tricks "The Invisible Pen" and "The Mathematician, the Psychic, and the Magician";
- Harold Stone, for giving freely (for decades) of his time and advice;
- Paul Swinford, for letting me use his trick "The Seekers";
- Byron Walker, for providing access to *The Whole Art and Mystery of Modern Gaming Fully Expos'd and Detected.*

Introduction

Recommendation: Before reading this amazing book be sure you have on hand a deck of cards, preferably a brand new one.

There are many curious ways in which mathematical theorems provide a basis for entertaining magic tricks, or, put another way, a raft of magic tricks depend for their eerie working on mathematical principles. Back in 1956 I had the pleasure of writing *Mathematics, Magic, and Mystery*, a book in which I collected for the first time a variety of such mystifications. Since then, hundreds of new and even more bewildering mathematical tricks have been invented, either by magicians fond of mathematics of by mathematicians fond of conjuring.

In no field of magic has mathematics played a greater role than in card magic. This is not only because cards bear numbers, colors, and suits, and have two sides, but also because cards can serve as counting devices; above all, they are counters easily randomized by shuffling. In recent decades a special kind of shuffle called a perfect shuffle, or (in the United States) a faro shuffle, has become essential for an enormous variety of mind-boggling tricks.

Although a few booklets on faro shuffling have been written exclusively for the magic fraternity, not until now has a "mathemagician" gathered in one place so many of the finest faro tricks, and probed to such depths the combinatorial theorems that lurk inside such conjuring.

Every now and then, to everyone's surprise, results emerging from recreational mathematics turn out to have practical applications. An outstanding recent instance is British physicist Roger Penrose's discovery of two simple polygons that tile the plane only in a nonperiodic fashion. Penrose had no inkling that a three-dimensional analog of his tiles would lead to the construction of what are called quasicrystals, now a hot field of research by crystallographers. He had constructed his tiles only for the fun of it. Something similar has happened with faro shuffling. As Brent Morris discloses in the latter part of his fascinating book, work done on faro theory for the fun of it has recently proved to be of great importance in computer science!

Magic Tricks, Card Shuffling, and Dynamic Computer Memories is essential reading for any magic buff who can faro shuffle or who wishes to acquire this unusual skill. The book will also be of great interest to computer scientists and to mathematicians working in the field of combinatorics. Needless to add, anyone can read it with enjoyment and profit who is curious about the art and mathematics of card magic, or about the unexpected application of perfect shuffles to the storage and retrieval of computer information.

I opened this introduction with a recommendation. Let me close with a warning: Faro shuffling can become addictive!

Martin Gardner
Hendersonville, NC
March 1997

1
The Perfect Shuffle

I was watching a guy kill time by shuffling cards. He cut the deck exactly in half, butted the halves together, and pushed slightly. The cards seemed to jump together, perfectly woven, every-other-one. He saw me watching him and said, "Come over here, and I'll show you a trick."

He had me choose a card and return it to the deck. The magician then reversed two cards side-by-side in the deck, and he asked me to cut the cards a couple of times. He then gave the cards a perfect shuffle, spread the deck, and pointed to a face-down card sandwiched between the two face-up cards. He asked me to turn over the sandwiched card. I did, but I couldn't believe my eyes. It was my card!

How did he do that?

I'm not sure how the idea came up of perfectly interlacing the halves of a deck. Maybe it was an enthusiast of some card game where the even cards won for the house and the odd cards won for the players. With a perfect shuffle, the cards from one half could be made to win for the house or the player, as the shuffler desired. Getting an edge in a card game can be a fierce motivator. Then again, a card player killing time might have decided to see if he could do a perfect shuffle—just for the sheer challenge of doing it. Stranger things have been mastered for no better reason. The simple, descriptive name *perfect shuffle* characterizes this action nicely.

The Origins of the Perfect Shuffle

The perfect shuffle traces its roots to an old card game, *faro*, and is still referred to by American magicians as the *faro shuffle*. The origins of the game of faro are unclear, but by 1726, *The Whole Art and Mystery of Modern Gaming* had a chapter devoted to "The Description of a Pharo-Bank, with the Expences and Attendants." [**6**, i] The game evolved in 18th-century France, and its name supposedly came from one of the cards of the pack at the time bearing the picture of a pharaoh [**82**, 78]. According to John Scarne, faro was the most popular gambling-house game from shortly after the Louisiana Purchase in 1803 until craps succeeded it in the early 1900s [**91**, 267]. The legendary lawman Wyatt Earp is said to have wanted to be a faro banker—that is, to run a faro game. Out West the game was advertised with a sign showing a tiger, and playing against a faro bank was known as "bucking the tiger." Its popularity has faded to the point that it is now virtually unknown, except at a few casinos trying to achieve an "old west flavor" by offering "old time" games.

  Faro is a simple game with little strategy, that allows players to lose slowly, perhaps even gracefully. Players bet on any of the thirteen values to win (suits don't count). In some versions players could bet on a card to lose by placing a small copper token on the wager, but we won't complicate our analysis by allowing this variation. (The copper tokens are thought to be the origin of the expression "copper a bet" or "cop a bet.") Bets on multiple cards are placed by putting the wager on the layout between cards (e.g., K-Q, 5-6, or 3-J) or in the center of four cards (e.g., 5-6-8-9 or 3-4-10-J). After bets are placed, the deck is shuffled and put face-up in a "shoe" (a box that allows cards to be dealt one at a time). See figure 1.

  At the start of a game, the first card in the shoe, seen by everyone, is called "soda" and doesn't count in play. Cards are then dealt in pairs; the first revealed is a loser and the second a winner. Soda is dealt aside to start a "winning" pile. The next card is a loser and is immediately dealt into a "losing" pile. The third card, now exposed in the shoe, is a winner. This completes a "turn"—one loser and one winner, and bets are settled. New bets are made, and another pair of cards is dealt. (The first card of the next pair dealt is the previous winner, and it is placed on the winning pile.)

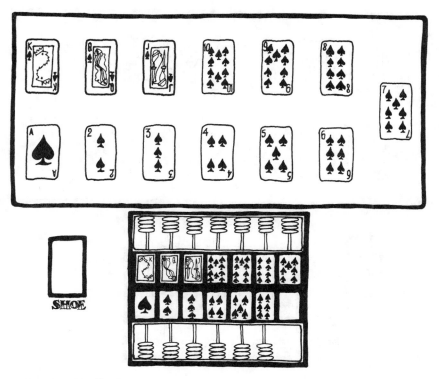

FIGURE 1. Faro betting layout and case-keeper as used in casinos.

Bets on the winning card are paid at 1 to 1. Bets on the losing cards are lost; bets on unplayed cards must remain (though sometimes house rules allow these bets to be moved). Bets on multiple cards are paid proportionately, e.g. a bet of 1 on K-Q is treated like one-half on K and one-half on Q, and if the winning and losing cards are the same, a "split," the house takes half the wagers on that card. Splits are what give the house some of its advantage, and we'll see later how dishonest dealers contrived to create splits with the perfect shuffle. A "case-keeper" or counting frame (usually thirteen wires, each with four beads) is used to keep track of the cards as they are played. A fully equipped faro bank requires a betting layout, a deck, a dealing shoe, and a case-keeper.

Let's follow the start of a simple game of faro, shown in figure 2. A shuffled deck is in the shoe, with the 8 of clubs as soda. The case-keeper records the one card displayed. Three bets have been placed: two chips on any of the sixes, one chip on the twos, and one chip on the jacks.

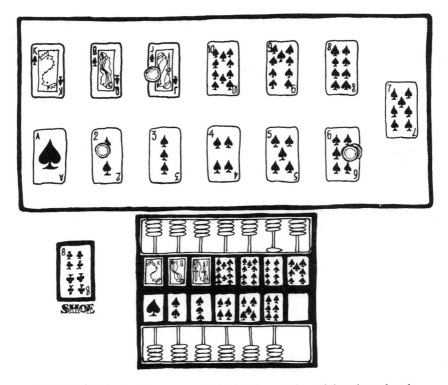

FIGURE 2.   Start of faro game with 8 of clubs as soda, and three bets placed.

Now the dealer starts a pile of cards with the soda, revealing the losing card of this turn—the 3 of diamonds. The soda, the 8 of clubs, neither wins nor loses, but the pile started with it will take the winners. A second pile is started with the loser, revealing the 2 of hearts—the winning card of this turn. Bets are now settled: There are no losers, as no one bet on the losing 3; the bet on the 2 is matched. The bets on the jacks and 6's must stay, and let's assume the bet and the winnings on the 2's also stays, as shown in figure 3.

The second turn reveals a losing jack of clubs and a winning 2 of spades. The wager on the jacks is lost and that on the 2's is matched again, bringing the total winnings there to four, as we see in figure 4. Again assume all bets ride.

The third turn reveals a split: The 6 of diamonds followed by the 6 of hearts. The house takes half of the wager on the sixes. This is shown in figure 5.

After 24 turns, three cards remain behind the face-up card in the shoe, which is the last winning card. The last three cards are known to the players be-

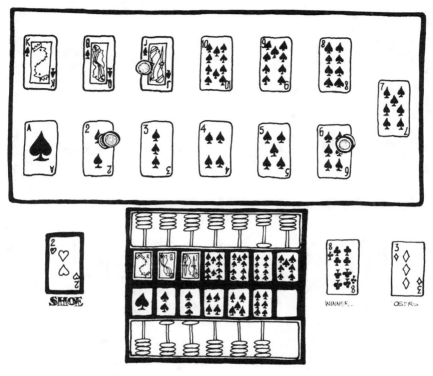

FIGURE 3. Cards, bets, and case-keeper after the first turn.

cause the case-keeper has tracked all cards played. Remaining bets are returned to the players, who can bet on the order of the last three cards. If all three cards are different, the house pays 4 to 1 for naming their order or "calling the turn" (the odds would be 5 to 1 in a fair game). If two of the cards are the same, a "cat hop," the house pays 1 to 1 for guessing their order (the odds would be 2 to 1 in a fair game). If all three cards are the same, there are no further bets.

Faro is a remarkably fair game for one still found in casinos (which are not known for paying out money on a regular basis). The game has been analyzed by such worthies as Abraham De Moivre [**16**], Richard Epstein [**24**], Leonhard Euler [**25**], Pierre de Montmart [**71**], and Edward Thorp [**100**].

There is anything but unanimity as to the house advantage, but Thorp's analysis seems to provide the best and most accurate understanding of the game. He analyzes bets on cards of value X by considering how many of those cards are left in the deck. [**100**, 842–844]

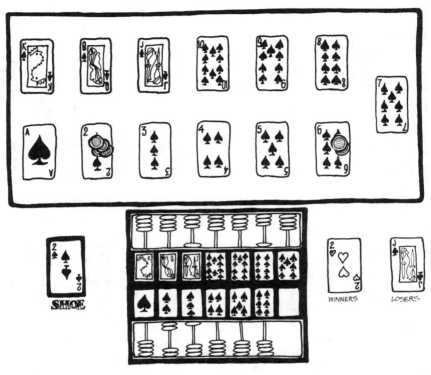

FIGURE 4.   Cards, bets, and case-keeper after the second turn.

If there is only one of card X left in the deck, then a wager on X is fair, and the player's expectation per unit bet is zero. In other words, the one card of value X is as likely to be a winner as a loser. Bets on turns in faro don't get any better than this.

If there are two cards of value X left in the deck, the most favorable situation for the player occurs when 49 cards are left (i.e., soda has been discarded and one turn has been played) and the bet is then resolved in one turn. In this case the expectation per unit bet is $-1/190 = -0.00526$. If there are less than 49 cards, the expectation is less.

If there are three cards of value X left in the deck, the most favorable situation for the player occurs when 51 cards are left (i.e., no turns have been played, but soda is known to be X) and the bet is resolved in one turn. The

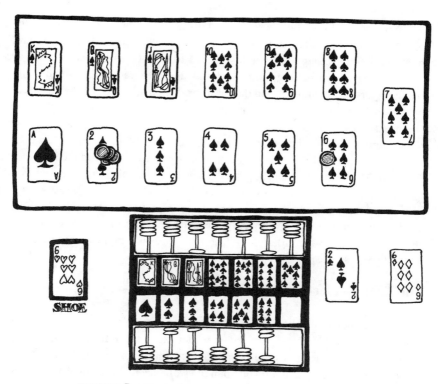

FIGURE 5. Cards, bets, and case-keeper after third turn.

expectation per unit bet is $-1/98 = -0.01020$. Anything else—more turns to resolution or fewer cards—yields a lower expectation. The worsening situation occurs because there are now more ways to get a split with three cards of value X.

Finally, if four cards of value X are left in the deck, the most favorable situation for the player occurs when 51 cards are left and the bet is resolved in one turn. The expectation per unit is $-3/194 = -0.01564$. Anything else yields a lower expectation.

Theorem 1.1. (Thorp) *The player who wishes to maximize the expectation per bet (conditioned on the bet being resolved) should, if he or she can alter bets after each turn, limit bets to values with a minimum nonzero number of cards remaining.* [**100**,844]

Note that while Thorp's theorem tells a player how to maximize the expectation per bet, that maximum is still a negative number. In playing faro, the best possible strategy lets you lose only 1 to 2 cents per $1 bet.

The analysis of betting on the order of the last three cards, which is *not* mandatory, is much simpler than that of turns. When all but the last three cards are played, the probability they are all alike is 0.0024, and the game ends. The probability there are three distinct cards is 0.8282. Calling the turn, that is, correctly guessing the order of three distinct cards, pays 4 to 1, and the player's expectation is -0.1667. The probability there is a pair in the last three is 0.1694. The payoff for a cat hop, that is, correctly guessing the order of a 2–1 split, is even, and the player's expectation is -0.3333. Thus, if a player always bets on calling the turn or on a cat hop, his expectation is -0.1948. In other words, you can lose money slowly, perhaps even gracefully, while betting on turns, but betting on the order of the last three cards is a rapid way to ruin.

The Faro Dealer's Shuffle

In 1726 the anonymous book, *The Whole Art and Mystery of Modern Gaming Fully Expos'd and Detected*, warned its readers against the follies of gambling and gave sound advice to would-be gamblers: "Shuffle the Cards well and take care they are not changed upon you, then bid Defiance to the Dealer." [**6**, 92–93] The book has a section on the game of basset, the immediate ancestor of faro. The author describes a perfect shuffle and alludes to a method used in faro.

> Suppose a, b, c, d—e, f, g, h to be certain Cards best known to yourself, it cannot be thought a difficult Task to joyn a to e, b to f, c to g, and d to h, and further to continue to 52 in the same Order.
>
> This was allow'd a fair Way to Shuffle.... Now if these Eight Letters (which I suppose Cards) were drawn thro' your Hand from top to bottom, as practis'd at *Faro*, then the Letters are changed to

a   h
b   g
c   f
d   e

[**6**, 91–92][1]

---

[1]The discovery of this earliest-known reference to the faro shuffle, and nearly every historical reference in this book, is due to Persi Diaconis, who has spent many years "pursuing the dovetail shuffle to its lair."

The first paragraph clearly describes what is now known as the *faro* or *perfect shuffle*: The deck is cut precisely in half and the cards interleaved every other one. This would bring together or "joyn" cards a and e, b and f, c and g, and d and h. If we let P represent the perfect shuffle, then

$$P[a, b, c, d, e, f, g, h] = [a, e, b, f, c, g, d, h].$$

The shuffle described in the next paragraph, however, is not a perfect shuffle. It seems to be a variant of what magicians call a *milk shuffle*: The thumb and fingers pinch the deck and draw off or "milk" the top and bottom cards as the rest of the cards are pulled away. Then milk the next pair of cards from the top and bottom and put them on top of the first pair. This is repeated until the entire deck is shuffled. If M represents the milk shuffle, then

$$M[a, b, c, d, e, f, g, h] = [d, e, c, f, b, g, a, h].$$

The permutation described in the book is a milk shuffle, but with successive pairs put underneath (instead of above) the previously milked cards. Connections between the milk shuffle and the perfect shuffle are discussed in [54] and [18].

Two things are significant here. By 1726 the perfect shuffle is described in print for the first time, and there is a specific type of shuffling known to be "practis'd at Faro." It will be 121 more years before these two ideas merge.

The first American description of the perfect shuffle is an obscure explanation in J. H. Green's 1847 book, *An Exposure of the Arts and Miseries of Gambling*, telling how to get certain cards into winning or losing positions in faro. The idea is to have these cards trimmed differently so they can be "stripped" out by pulling on the side of the deck. Such a deck is called a "stripper deck," and versions are sold today in magic stores. The commercial cards are tapered slightly to form trapezoids, with the short edges parallel to each other. If a card is reversed in the deck, its wide end protrudes from the narrow ends of the rest of the cards, and can be easily stripped out.

After stripping out the desired cards and perfectly shuffling or "running in" the cards, specific values can be put in winning or losing—odd or even—positions. Here's the rather poor way Green described the process in his section on "Deceptions Used in the Game of Faro."

> At other times, the odd cards, namely, 1, 3, 5, 7, 9, jack, and king are trimmed differently from the remainder of the cards, and their ends reversed. This cheat is of late introduction, and not so generally known, and is often performed on that account. They are then pulled and run in, an odd against

an even, and they tell by the different sizes of cards: these are trimmed, as I have before spoken of. [**34**, 136]

Green's explanation may be hard to follow, but it is the first in print to connect the perfect shuffle with the game faro. In 1860 a much better description of the shuffle appeared in another anonymous book, *A Grand Exposé of the Science of Gambling*. The author boasts that Green would now be an easy victim to professional gamblers.

> Several years since a reformed gamester, by the name of Green, placed a work before the public, purporting to be an *exposé* of the mysteries of gaming. This book doubtless produced a very desirable effect upon the public mind, but it would not be of much value at the present day, as such skillful improvements have been made during the past ten or fifteen years in gaming, that I doubt not Mr. Green himself would now prove an easy victim to the professional gamester, provided he has not been engaged at gaming since he issued his book on the subject. [**4**, 4]

The use of the shuffle remained the same as with J. H. Green: to cheat at faro. After stripping out predetermined cards, the dealer perfectly shuffles the halves together so the desired cards are either in winning or losing positions. This description sounds like the way magicians perform the perfect shuffle today.

> After shuffling them a sufficient length of time, the dealer suddenly, and with a slight movement of his hands, pulls or strips the deck as above described; then taking one half the deck in one hand, and the other half in the other hand, and placing the ends together, runs them in, thereby displacing every card in the deck by the process of running them. [**4**, 6]

In 1894, *Koschitz's Manual of Useful Information* gave a much better description of the perfect shuffle, which it called "butting-in." A determined shuffler might even be able to do the shuffle from this description.

> If one half of a pack be taken in each hand and their ends pressed slightly together, the cards of one half could readily be made to enter one over another in the other half, interlapping each other by the momentary springing of the pressure given, a mode of shuffling which, in gambling parlance, is called butting in. [**5**, 27–28]

In 1894 the perfect shuffle also was illustrated for the first time and given the name of "The Faro Dealer's Shuffle" in John Nevil Maskelyne's *Sharps and Flats*. Maskelyne was a well-known English magician, and his book was subtitled "a complete revelation of the secrets of cheating at games of chance and skill." In addition to using the perfect shuffle to put certain cards in winning or losing positions, he explained how to shuffle "splits" into the deck. This was accomplished as the dealer threw "the cards carelessly into two heaps, instead of making two even piles." [**67**, 203] As he gathered up the two piles he looked for pairs of cards and casually placed them so that one of each pair was in the same position in each pile. All that was left was to bring the pairs together. Maskelyne's "laterally reciprocating motion" was the best description of the shuffle to date.

> This is accomplished by means of what is called the "faro dealer's shuffle." It must not be thought that this manipulative device is essentially a trick for cheating; on the contrary, it is an exceedingly fair and honest shuffle, provided that there has been no previous arrangement of the cards. By its use, a pack which has been divided into two equal portions may have all the cards of one half placed alternately with those of the other half at one operation. In faro, the manner of dealing the cards necessarily divides them into two equal parts. This being the case, they are taken up by the dealer, one in each hand. Holding them by the ends, he presses the two halves together so as to bend them somewhat after the manner shown in figure 6, in the position "A." The halves are now "wriggled" from side to side in opposite directions with what would be called in mechanism a "laterally reciprocating motion." This causes the cards to fly up one by one, from either side alternately, as indicated in the figure at "B." Thus it is evident that those cards which have been placed, with malice aforethought, in corresponding positions in the two piles, will come together in a shuffle of this kind, and form splits. [**67**, 204]

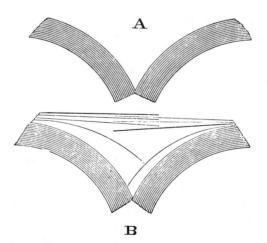

FIGURE 6. The "faro dealer's shuffle." John Nevil Maskelyne, *Sharps and Flats*, 2nd ed. (London: Longmans, Green and Co., 1894), p. 205.

With Maskelyne's description, the perfect shuffle emerges into its modern form. It is now inextricably connected with faro. A century later, American magicians still refer to it as the "faro shuffle," though British magicians prefer the term "weave shuffle." Maskelyne's method of shuffling has evolved to the point that there is little room for improvement of the technique. Edward Marlo's 1958 book, *The Faro Shuffle*, explains the shuffle with diagrams quite similar to Maskelyne's. [**63**, 3] (See Appendix 2 for my description of the faro shuffle.) Maskelyne went on to summarize the skill needed to perform and the value of the shuffle.

> This shuffle is a very difficult one to learn; but with practice and patience it can be accomplished, and the cards can be made to fly up alternately, without any chance of failure. A dealer, skilled in the devices we have just touched upon, can put up four or five splits in one deal, if he think it advisable so to do. [**67**, 205]

A Mathematical Model of the Perfect Shuffle

The great appeal of the perfect shuffle to faro dealers and magicians is its regularity. After a perfect shuffle, the dealer knows exactly what happens to each card, and the same thing happens with every shuffle. Precision of this sort usually indicates an underlying mathematical model, and this is certainly the case here.

Before we develop the model, we observe there are two types of perfect shuffles: The top card is left on top or "out" after the shuffle, or it is shuffled "in." Consider a deck of ten cards numbered 0 to 9 from the top. If $\mathbf{O}$ represents the shuffle that leaves the top card *out* and $\mathbf{I}$ represents the shuffle that leaves the top card *in*, then

$$\mathbf{O}[0, 1, 2, 3, 4, 5, 6, 7, 8, 9] = [0, 5, 1, 6, 2, 7, 3, 8, 4, 9]$$

and

$$\mathbf{I}[0, 1, 2, 3, 4, 5, 6, 7, 8, 9] = [5, 0, 6, 1, 7, 2, 8, 3, 9, 4].$$

This terminology of *out-shuffles* and *in-shuffles* was coined by British magician and computer scientist Alex Elmsley, a pioneer researcher into the mathematics of the perfect shuffle.

> Some while ago, when making notes of tricks using the weave shuffle, I started using the abbreviations "in-shuffle" and "out-shuffle" for the two ways of weaving a pack containing an even number of cards. With an out-

shuffle the top and bottom cards remain outside the rest of the pack; with an in-shuffle they go inside the pack (with one card above and below them respectively). [**22**]

It is fairly easy to see in a deck of $N = 2n$ cards, that after one out-shuffle, the card in position p is moved to position $\mathbf{O}(p) \equiv 2p \pmod{N - 1}$, $0 \le p < N - 1$, and $\mathbf{O}(N - 1) = N - 1$. After an in-shuffle, $\mathbf{I}(p) \equiv 2p + 1 \pmod{N + 1}$.

To get a full definition for perfect shuffles we need to consider what happens when the deck has $N = 2n - 1$ cards. Magicians call perfect shuffles on odd decks "straddle shuffles" because the larger packet straddles the smaller one.

To define an out-shuffle on an odd deck, recall that after an out-shuffle on an even deck, the bottom card remains fixed, $\mathbf{O}(N - 1) = N - 1$. Thus removing it doesn't affect the shuffle one way or another; the top packet of n cards is shuffled into the bottom packet of $n - 1$. For an odd deck of 9 cards,

$$\mathbf{O}[0, 1, 2, 3, 4, 5, 6, 7, 8] = [0, 5, 1, 6, 2, 7, 3, 8, 4].$$

In a parallel manner, we can perform an in-shuffle with $n - 1$ cards in the top packet and n in the bottom. Thus for a deck of 9 cards

$$\mathbf{I}[0, 1, 2, 3, 4, 5, 6, 7, 8] = [4, 0, 5, 1, 6, 2, 7, 3, 8].$$

The formulas turn out to be "nicer" for odd decks of $N = 2n - 1$ cards: The modulus is the same for out- and in-shuffles, N. See figure 7.

Definition. The *out perfect shuffle* or *out faro shuffle* or *out-shuffle* on a deck of N cards, $0, \ldots, N - 1$, is the permutation that moves the card in position p to position $\mathbf{O}(p)$ where

$$\mathbf{O}(p) \equiv 2p \begin{cases} \pmod{N - 1} & \text{for } N \text{ even and } 0 \le p < N - 1 \\ \pmod{N} & \text{for } N \text{ odd and all } p, \end{cases}$$

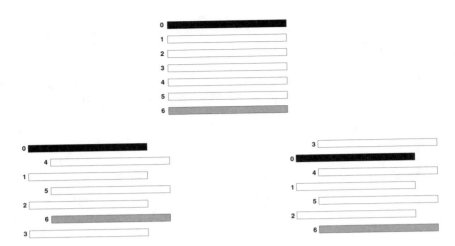

FIGURE 7. Perfect shuffles on odd decks are called "straddle shuffles" by magicians because the larger packet straddles the smaller one. An out- and in-shuffle are shown on the left and right respectively.

and

$$\mathbf{O}(N - 1) = N - 1 \quad \text{for } N \text{ even.}$$

Definition. The *in perfect shuffle* or *in faro shuffle* or *in-shuffle* on a deck of N cards, $0, \ldots, N - 1$, is the permutation that moves the card in position p to position $\mathbf{I}(p)$ where

$$\mathbf{I}(p) \equiv 2p + 1 \begin{cases} (\mod N + 1) & \text{for } N \text{ even} \\ (\mod N) & \text{for } N \text{ odd.} \end{cases}$$

This definition of an out-shuffle for odd decks was first given by Paul Lévy [**50**, 423], though he numbered his deck from $1, \ldots, N$ rather than $0, \ldots, N - 1$. He discusses the permutation $\mathbf{O}$ in relation to the milk shuffle M, without calling either permutation a shuffle or indicating any connection with or motivation from cards. In a subsequent paper [**51**], however, he explains the card shuffling model for M, but never connects $\mathbf{O}$ with cards. Lévy's original motivation for studying shuffle permutations was not revealed until his autobiography [**52**, 151–153], when he explained that as a young boy a German lieutenant showed him a card trick with the milk shuffle. Almost fifty years after that he began studying the trick, from which resulted several papers. A nice description of Lévy's work on the cycle structure of $\mathbf{O}$ is in [**18**, 190–191].

The Stay-Stack Principle

The faro shuffle has been used by magicians for some of its physical properties, with no apparent realization of its mathematical attributes. Starting a perfect shuffle and pushing the cards partially together lets the manipulator spread them in a spectacular double fan. [**99**, 124–127] and [**37**, 1:16–19] Another clever principle, which can be discovered without any mathematics, is known as the "stay-stack," invented by J. Russell Duck. [**20**] Duck's idea was to prearrange or "stack" an even deck so that cards matching in color and value are located at k from the top and k from the bottom, for all k. Such an arrangement has "central symmetry."

Any number of faro shuffles, either out or in, maintains the bottom half of a stay-stacked deck as a "mirror image" of the top. Numbering the deck $0, \ldots, N - 1$, the cards k from the top and bottom are in positions $k - 1$ and $N - k$. An in-shuffle moves the cards as follows, with all congruences modulo $N + 1$: $\mathbf{I}(k - 1) \equiv 2k - 1$, and $\mathbf{I}(N - k) \equiv 2N - 2k + 1 \equiv N - 2k$. Similarly for an out-shuffle, with congruences modulo $N - 1$, $\mathbf{O}(k - 1) \equiv 2k - 2$, and $\mathbf{O}(N - k) \equiv 2N - 2k \equiv N - 2k - 1$. Thus $\mathbf{I}(k - 1) + \mathbf{I}(N - k) = N - 1$, and $\mathbf{O}(k - 1) + \mathbf{O}(N - k) = N - 1$.

Paul Swinford moved the idea of the stay stack a giant step forward by using an odd deck with a joker in the middle [**98**, 23–25]. Now any number of in- and out-shuffles plus cuts, maintains central symmetry around the joker.

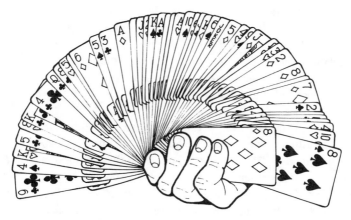

FIGURE 8. A double fan from two perfectly shuffled decks of cards.

That is to say, the nth card "before" the joker mirrors the nth card "after" the joker (where counting is "around the deck"). To see this, assume an odd stay-stacked deck has been cut so the joker is in some position j, and cards $j + k$ and $j - k$ match in color and value for all k. All congruences are now modulo N. For the out-shuffle,

$$\mathbf{O}(j) \equiv 2j, \qquad \mathbf{O}(j - k) \equiv 2j - 2k, \qquad \text{and} \qquad \mathbf{O}(j + k) \equiv 2j + 2k,$$

$$\mathbf{I}(j) \equiv 2j + 1, \quad \mathbf{I}(j - k) \equiv 2j - 2k + 1, \quad \text{and} \quad \mathbf{I}(j + k) \equiv 2j + 2k + 1.$$

Also, if the deck is cut again so the joker shifts s positions, then $\mathbf{C}(j) \equiv j + s$, $\mathbf{C}(j + k) \equiv j + k + s$, and $\mathbf{C}(j - k) = j - k + s$.

Here's how Swinford's odd stay stack can be used. The magician brings out the stacked deck and offers it to be cut, then does a few faro shuffles, and lets the deck be cut several more times. Before doing any magic, the magician says, "Oh, I have to take the joker out for this trick," and runs through the cards face up until he finds the joker. (Note that the cards on either side of the joker match in value and color, as do the cards k above and below the joker.) After the joker is removed, the deck is reassembled so the cards on either side of the joker are now the top and bottom cards of the deck. An even stay stack now results, and any number of shuffles (but no cuts) will maintain the stack.

Trick 1.2. ("The Seekers," Paul Swinford) The magician starts a faro shuffle but doesn't push the halves together to complete it. Holding one half in his hand with the other half sticking up, the magician riffles the corner of the top half and has a spectator say "Stop" to select a card. The spectator peeks at the card and remembers it. The magician pulls out the top half, reassembles the deck, and cuts it once. He or she attempts to cut to the spectator's card but fails. The card above and the card below the cut are reversed on top of the deck, and the spectator cuts them somewhere into the center. The magician faro shuffles the deck once and spreads it on the table. The chosen card is found sandwiched between the two reversed cards.

Explanation. This, to my mind, is the best faro shuffle trick around. It is Paul Swinford's brilliant invention. "The Seekers" was first published in his 1968 book, *Faro Fantasy* [97], and is used here with his permission. Paul Le Paul used an incomplete faro shuffle for his trick, "The Gymnastic Aces," but he used only physical properties of the shuffle, not mathematical. [49] A decade later Edward Marlo explored the possibilities of using an incomplete faro to control the placement of a chosen card. [58, 64, 65, 66] Some of Marlo's work involved the spectator peeking at a card in the top half and the performer holding a break in the bottom half, much like Swinford does for "The Seekers." Marlo, like Le Paul, never used any mathematical properties of the incomplete faro for his controls.

Paul Swinford, in the best tradition of experimental science, made his discoveries playing in "the laboratory." While experimenting with the incomplete faro, he noticed he could control a chosen card to the exact middle of the deck. With the addition of a joker to produce an odd-sized deck, he found he could cut the deck and bring the sandwiched cards somewhere in the middle. The result was "The Seekers," which succeeds both mathematically and theatrically. There are other faro shuffle tricks that involve more sophisticated mathematics, but they just don't leave the audience as stunned as this one. It's so good that it alone makes learning the faro shuffle worthwhile.

(In the explanation, the magician's dialogue or "patter" is shown in ***bold italic*** type.)

Let me show you an example of what gamblers call "card control." I'll cut off half the deck, because that's all I need, and I'll check that it's exact by shuffling the halves together.

Using an odd deck, perform a modified incomplete in-shuffle. (The shuffle is "modified" because two cards are together on the face and "incomplete" because the two halves are not pushed together; see figure 9.) I usually use a

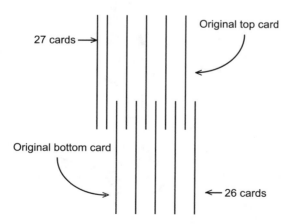

27 cards →

Original top card

Original bottom card

← 26 cards

FIGURE 9. The modified incomplete in-shuffle used in "The Seekers."

joker to make a deck of fifty-three, but you can just as easily leave a card in the case and use a deck of fifty-one. Assume you're using a deck of fifty-three. Cut off twenty-seven as if you were going to do a straddle out-shuffle, but in-shuffle the larger packet into the smaller. This will leave two cards together on the face of the deck.

You can choose any card you want from the top half of the deck. Just say "Stop" while I riffle the corners. Take a peek at your card and remember it.

Riffle the top half of the partially shuffled packets with the index finger and ask the spectator to say "Stop" to choose a card. Be sure to hold the cards so their faces are towards the spectator. As you hold the deck in your left hand, the tip of your left little finger will naturally rest against the outer edge of the cards. When the spectator says "Stop," pull back the corner of the chosen card with the right index finger. Let the spectator peek at the card. As you pull back the card, pull it back enough to cause a separation or "break" in the bottom

FIGURE 10.   As the right index finger pulls back the corner of the chosen card, the left little finger is inserted in the "break" in the bottom half.

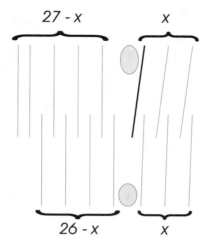

FIGURE 11.   The right index finger is shown above and the left little finger below (in the "break"). The chosen card is highlighted.

half. Carefully insert the tip of your left little finger in the break before you remove your right index finger from the top half, as in figure 10.

Recapitulation. At this point the spectator's card is somewhere in the top half of the deck, x cards from the top (figure 11). You don't know where it's located, but you have a break held in the bottom packet exactly x cards from the top. You're now ready to place the chosen card just where you want it to be.

Your card is somewhere in this half of the deck.

As you say this, pull out or "unweave" the top half of the deck and put it on the table. Pause for just a second to let it sink in that the chosen card is sitting somewhere in the half of the deck on the table. Continue to hold the little-finger break in the cards in your left hand.

Let's put the deck back together and see if I can find your card with my card control.

Pick up the cards on the table and put them on top of those in your left hand. Cut the deck at the little-finger break, put the $26 - x$ cards from below the break on top, and place the reassembled deck on the table. The hard work is over; all you need now is one faro shuffle and a little showmanship.

Recapitulation. Let's reconstruct where we are after the deck has been reassembled. Before reassembly, half of the deck was in your left hand and the other half was on the table. The bottom half, in your hand, had 26 cards separated by your little finger into groups of $26 - x$ and x. The top half, on the table, had 27 cards with the chosen card x from the top. When you reassembled the deck, the chosen card is x from the top, and you held a break $26 - x$ from the bottom.

$$\text{original top part} \begin{cases} x \\ 27 - x \end{cases} \qquad \text{current top}$$

$$\text{original bottom part} \begin{cases} \dfrac{x}{26 - x} \qquad \text{current bottom} \end{cases}$$

This shows the deck as held in your hand after reassembly, and before cutting the bottom $26 - x$ cards to the top. The break is indicated by the heavy line.

$$\begin{array}{ll} 26 - x & \text{current top} \\ \text{original top part}\begin{cases} x \\ 27 - x \end{cases} & \\ x & \text{current bottom} \end{array}$$

This shows the deck as held in your hand after cutting the bottom $26 - x$ cards to the top.

When the bottom $26 - x$ cards are cut to the top, the chosen card is moved from x from the top to 26 from the top. Its position changes from $x - 1$ to 25. (Remember the positions in the deck are numbered from 0 to $N - 1$.) The deck is now sitting on the table.

I'll cut to your card to find it.

Cut off and hold in the right hand about the top two-thirds of the deck, without showing the bottom card. What's important is that you cut off more than half, because the chosen card is 26 from the top, and you want the chosen card in the top half.

There, how's that? Well, I guess since I haven't shown it to you yet, it's not too impressive.

Of course, the audience won't be impressed, since you haven't shown them anything!

Let me make a confession about how this trick works. I really don't have perfect control of the deck. I've used a subtle ambiguity in English to double my chances of getting your card. When I said, "I'll cut to your card," there are really two cards that satisfy that statement. It could be this one.

Point upward with your left index finger to the bottom card of the portion you cut off.

Or it could be this one.

Point to the top card of those on the table (the bottom part).

Let's see how I did. Is this your card?

Show the card on the bottom of the packet you're holding in your right hand. When the spectator says, "No," frown a little, place the card face up to the side, and put these cut-off cards on the table near the bottom part of the deck.

Then this must be your card.

Turn over the top card of the bottom part as you say this and show it to the spectator. The spectator will say you've failed again.

Hmm, this usually doesn't happen. Let's see what I can do to get out of this fix.

Put the two wrong cards face up on top of the thicker part (the original top of the deck), and put the smaller part (the original bottom of the deck) on top of this. You now have two face-up cards together somewhere in the deck, and you're ready for the grand climax.

Cut the cards for me, please—anywhere you like. Since this is a perfect mess, I'll give the cards a perfect shuffle. Do you remember those two wrong cards that were left face up together in the deck? Your cut put them some where in the deck, and my perfect shuffle put one card between them. Wouldn't it be strange if your card sense put my wrong cards in just the right place to find your card?

Let the spectator make any number of straight cuts, and perform an out- or in-shuffle. Spread the cards to show the chosen card sandwiched between the two reversed cards. Don't get carried away with having the spectator cut the cards, or you can lose some of the drama. I usually ask for only one cut, but two wouldn't be bad. If they're not satisfied with two free cuts, then seven won't make them feel much better. And it will detract from the climax you're building.

Recapitulation. The chosen card was in position 25 in the thicker part cut off the top of the deck. The two wrong cards were put face-up on these cards in positions 0 and 1, moving the chosen card to position 27. The cards from the table were put on top of the two face-up cards, and the spectator's cuts put some more cards, say a total of x, on top of these. This moves the three cards to positions x, $x + 1$, and $x + 27$, all modulo 53. After an out-shuffle, the cards are moved to positions $2x$, $2x + 2$, and $2x + 54 \equiv 2x + 1$, all modulo 53. After an in-shuffle, the cards are in positions $2x + 1$, $2x + 3$, and $2x + 55 \equiv 2x + 2$, again all modulo 53. In other words, either an in-shuffle or an out-shuffle bring the three cards together, with the chosen card sandwiched between the two face-up ones.

Why don't you check out the card in the middle?

Let the spectator remove the sandwiched card and turn it over. If all goes well (and it usually does for me) there will be a small gasp as the chosen card is revealed.

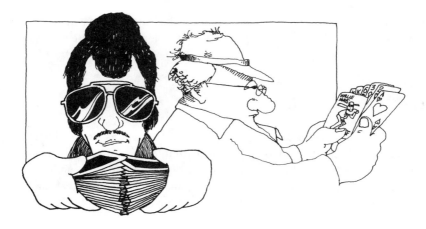

2
The Order of Shuffles

I was in the audience watching a magician do a card trick. A deck was tossed into the crowd, and someone named Susan caught it. She shuffled the deck several times and chose a card.

With a flourish, the magician pulled out of his pocket an "invisible pen." He told Susan to "sign" the back of her card (which was a little odd, since the pen was "invisible") and then to show the card to the rest of the spectators.

When this was done, the magician had the deck reassembled and shuffled several times. He returned the deck to its case.

Once more he tossed the deck into the audience. A woman named Liz caught it and was asked to toss it to someone else; a fellow named Dave caught it. The magician asked Dave to spell his name aloud, dealing one card for each letter, D-A-V-E. When he finished, the magician asked Dave to turn over the card he stopped at, which was Susan's card. Tumultuous applause spontaneously erupted accompanied by wild cheering.

How did he do that?

One simple fact spurred my interest in studying the perfect shuffle: eight perfect shuffles restore a deck of fifty-two cards to its original order. I was about sixteen when a friend told me this, and I simply didn't believe it. It was counterintuitive that shuffling a deck of cards only eight times had no effect on it. In the best fashion of scientific discovery, I set out either to humiliate my obviously foolish friend or to establish conclusively this amazing fact. My friend was not as foolish as I thought, and this fact leads us to more interesting results.

The Order of Shuffles

We shall establish the fact that eight perfect shuffles restores a normal deck as well as the general rule for finding how many shuffles restore a deck of any size to its original order. First, however, we need a definition and a few preliminary theorems. (This is a math book, after all, and math books seldom get directly to any result—a solid foundation always comes first.)

Definition. *The order of a shuffle* for the shuffle **S** on a deck of N cards, $o(\mathbf{S}, N)$, is the smallest number of shuffles that returns the deck to its original order.

The result with which my friend impressed me can be restated in this notation as $o(\mathbf{O}, 52) = 8$. The preliminary theorems are pretty simple, and the most effective proofs are visual "proofs without words." A little thought on the diagrams should do the trick. Be sure to note that certain cards have been highlighted—the top, bottom, or next-to-the-top cards, for example.

Theorem 2.1. *The order of an out-shuffle, $\mathbf{O}$, for a deck of $2n - 1$ cards is the same as the order of an out-shuffle for a deck of $2n$ cards: $o(\mathbf{O}, 2n - 1) = o(\mathbf{O}, 2n)$.*

Proof with very few words. An out-shuffle with $2n - 1$ cards is indistinguishable from an out-shuffle with $2n$ cards whose bottom card is "invisible."

Proof without words.

Theorem 2.2. $o(\mathbf{I}, 2n - 1) = o(\mathbf{O}, 2n)$.

Proof with very few words. This proof is just like that of Theorem 2.1. An in-shuffle with $2n - 1$ cards is indistinguishable from an out-shuffle with $2n$ cards whose top card is "invisible."

Corollary 2.3. $o(\mathbf{I}, 2n - 1) = o(\mathbf{O}, 2n - 1)$.

Theorem 2.4. $o(\mathbf{I}, 2n - 2) = o(\mathbf{O}, 2n)$.

Proof without words.

A summary of these results shows all shuffle orders can be expressed in terms of the order of an out-shuffle for a deck of $2n$, $o(\mathbf{O}, 2n)$:

$$o(\mathbf{O}, 2n - 1) = o(\mathbf{O}, 2n);$$

$$o(\mathbf{I}, 2n - 1) = o(\mathbf{O}, 2n);$$

$$o(\mathbf{I}, 2n - 2) = o(\mathbf{O}, 2n).$$

If we can compute $o(\mathbf{O}, 2n)$, we will have numerical answers to all other questions about the orders of perfect shuffles.

In a deck of $N = 2n$ cards,

$$\mathbf{O}(p) \equiv 2p \bmod (2n - 1),$$

and

$$\mathbf{O}^k(p) \equiv 2^k p \bmod (2n - 1).$$

The order of the out-shuffle, $o(\mathbf{O}, 2n)$, is the smallest k such that for $0 \leq p < 2n$

$$\mathbf{O}^k(p) \equiv p \bmod (2n - 1),$$

or, equivalently, k is the order of $2 \bmod (2n - 1)$:

$$2^k \equiv 1 \bmod (2n - 1).$$

[In general, if gcd $(u, N) = 1$, the "order of $u \bmod (N)$" is the smallest k such that $u^k \equiv 1 \bmod (N)$].

Using Euler's φ function from elementary number theory, we know that for gcd $(a, N) = 1$, $a^{\varphi(N)} \equiv 1 \pmod{N}$, and thus

$$o(\mathbf{O}, 2n) \mid \varphi(2n - 1).$$

This tells us nearly all there is to know about $o(\mathbf{O}, 2n)$.[2] We can give this divisibility condition of $o(\mathbf{O}, 2n)$, but the only strategy for computing is to try all divisors of $\varphi(2n - 1)$. There is just no simple way to compute $o(\mathbf{O}, 2n)$ as there is no simple way to compute $\varphi(2n - 1)$, because computing this function ultimately involves factoring $2n - 1$. Factoring remains a computationally difficult task, infeasible for numbers over a few hundred digits.

Table 1 gives several values of $o(\mathbf{O}, N)$ and $o(\mathbf{I}, N)$, and Appendix 1 gives all values of $o(\mathbf{O}, N)$ and $o(\mathbf{I}, N)$ for N up to 200. (This is surely the largest deck of cards anyone will try to shuffle perfectly!)

The answer to how many shuffles are needed to return a deck of 52 to its original order has been known since at least the early 1900s. The earliest reference appeared in 1915 by S. Victor Innis, but it would not be surprising if other card players had discovered this earlier still. "By riffling a pack eight times with this riffle, each card is brought back to its original position." [**41**, 13] Five years later Charles T. Jordan showed the evidence of experimentation

---

[2]We can show that if $8 \mid N$, and $(a, n) \equiv 1$, then $a^{1/2\varphi(N)} \equiv 1 \pmod{N}$, and $o(\mathbf{O}, 2n) \mid 1/2\varphi(N)$.

TABLE 1   The Orders of Perfect Shuffles

N	$o(\mathbf{O}, N)$	$o(\mathbf{I}, N)$	N	$o(\mathbf{O}, N)$	$o(\mathbf{I}, N)$
2	1	2	13	12	12
3	2	2	14	12	4
4	2	4	15	4	4
5	4	4	16	4	8
6	4	3	17	8	8
7	3	3	18	8	18
8	3	6	⋮	⋮	⋮
9	6	6	51	8	8
10	6	10	52	8	52
11	10	10	53	52	52
12	10	12	54	52	20

by perfectly shuffling different sized decks. "It may interest the reader to know that a piquet pack of 32 cards, pre-arranged, may be brought back to perfect order by shuffling it with absolute precision five times; but it requires considerably more skill to handle that many cards without mishap. Mr. [T. Nelson] Downs, however, can handle a full pack of 52 cards with the degree of dexterity necessary to restore its original order." [**43**, 34] T. Nelson Downs was a well-known professional magician and a contemporary of Harry Houdini. Downs did not use Maskelyne's faro-shuffle technique of butting together the ends of the two halves of a deck. Rather, Downs achieved a perfect shuffle by riffling the two halves together. Here's how he described the shuffle in a circa 1923 letter to fellow magician, Edward G. "Tex" McGuire. "Eight complete perfect dovetail shuffles, breaking pack exactly in center; that is, cutting off just 26 cards each time and dropping cards from each half alternately brings the pack to its original order." [**11**, 64] John Northern Hilliard used some hyperbole in describing the skill of perfect-shufflers. "I know perhaps a dozen card handlers who can take a new pack out of its case and with eight perfect riffles bring the deck back to its original order.... But such delicacy and absolute precision is not for the many. It is the poetry of card work—and poets are always few in any art." [**37**, 1:18]

In a 1924 letter to McGuire, Downs wrote about a Thedeford, Nebraska cattle rancher named Fred Black, who was another pioneer shuffle researcher. "Black writes me he hopes to see me in next 30 days—says he cannot explain his shuffles on paper." [**11**, 79] Black had achieved some fame in Ripley's "Believe It Or Not," as "the only man who could take a shuffled pack of cards, deal out four Bridge hands, then gather the cards together, riffle shuffle, have the cards cut, shift the cut and deal out the identical hands to the same players in less than a minute! Fred Black also supplied the Faro tables to T. Nelson Downs which appeared in *Expert Card Technique*." [**29**, 86] These "Faro tables" show the cycle decomposition of the out-shuffle on a deck of 52: two fixed points (the top and bottom cards), one transposition, and six 8-cycles. [**40**, 146–147] Despite his great shuffling skill, Black's discoveries were empirical; he seemed to have no idea of the mathematics behind the permutations. In fact, his magic was limited to "shuffling-up." As Downs said, "This is all he does except a couple or three fine shifts." [**11**, 67]

The Product of Shuffles

Magicians, card cheats, and mathematicians would never be satisfied with doing just one type of shuffle. Each has motives for further research: fooling the audience, fleecing the marks, or impressing the dean. (It could be argued these are equivalent activities.) Regardless of the motives, there are some interesting results about the interaction of out- and in-shuffles. As before, we start with a definition.

Definition. [3] The *function* δ is a mapping from perfect shuffles to binary values:

$$\delta : \{\mathbf{O}, \mathbf{I}\} \longrightarrow \{0, 1\}.$$

$$\delta(\mathbf{S}) = \begin{cases} 0, & \mathbf{S} = \text{Out}, \\ 1, & \mathbf{S} = \text{In}. \end{cases}$$

When describing the movement of shuffled cards in an even deck the modulus is different for out-shuffles $(N-1)$ and in-shuffles $(N+1)$, respectively. In an odd deck, however, the modulus is constant, making the formulas more tractable. Recall the definitions of the out- and in-shuffles in an odd deck of N cards:

$$\mathbf{O}(p) \equiv 2p \bmod(N)$$

---

[3]There seems to be an unwritten requirement among mathematicians to use Greek letters for notation, perhaps to give their work an air of mystery (as if mathematics needs to be more mysterious). Rather than fight so well-established a custom, I slavishly have used Greek letters— δ, φ, and Σ in this chapter alone.

and

$$\mathbf{I}(p) \equiv 2p + 1 \bmod(N).$$

We can now create a unified definition of perfect shuffles using $\delta(\mathbf{S})$, where $\mathbf{S}$ represents either shuffle.

Definition. A *perfect shuffle* $\mathbf{S}$, either Out or In, on an odd deck of $N = 2n - 1$ cards, moves the card in position p as follows:

$$\mathbf{S}(p) \equiv 2p + \delta(\mathbf{S}) \bmod(N).$$

We now have the notation to state and prove one of the most surprising and useful theorems of perfect shuffling, at least to magicians and electrical engineers. The theorem was first stated in [**22**] and its proof was first published in [**73**].

Theorem 2.5. (Morris) *The Fundamental Theorem of Faro Shuffling in Odd Decks. In an odd deck of $N = 2n - 1$ cards, a sequence of k perfect shuffles, $\mathbf{S}_i$, $1 \le i \le k$, moves the card in position p as follows:*

$$\mathbf{S}_k \cdots \mathbf{S}_2 \mathbf{S}_1(p) \equiv 2^k p + \sum_{i=1}^{k} 2^{k-i} \delta(\mathbf{S}_i) \bmod(N).$$

Proof. For a single shuffle, $\mathbf{S}_1$, on a deck of N cards,

$$\mathbf{S}_1(p) \equiv 2p + \delta(\mathbf{S}_1) \bmod(N).$$

After two shuffles,

$$\mathbf{S}_2 \mathbf{S}_1(p) \equiv 2^2 p + \delta(\mathbf{S}_1)2 + \delta(\mathbf{S}_2) \bmod(N).$$

After three shuffles,

$$\mathbf{S}_3 \mathbf{S}_2 \mathbf{S}_1(p) \equiv 2^3 p + \delta(\mathbf{S}_1)2^2 + \delta(\mathbf{S}_2)2 + \delta(\mathbf{S}_3) \bmod(N).$$

And in general, after k shuffles,

$$\mathbf{S}_k \cdots \mathbf{S}_2 \mathbf{S}_1(p) \equiv 2^k p + \delta(\mathbf{S}_1)2^{k-1} + \cdots + \delta(\mathbf{S}_{k-1})2^1 + \delta(\mathbf{S}_k)2^0 \bmod(N)$$

$$\equiv 2^k p + \sum_{i=1}^{k} 2^{k-i} \delta(\mathbf{S}_i) \bmod N \qquad \square$$

Note that the sum $\sum_{i=1}^{k} 2^{k-i} \delta(S_i)$ is a function solely of the shuffles $S_1, \ldots, S_k$ and is independent of p. We call this quantity the *shift* of a sequence of shuffles. The shift can be easily calculated by viewing the shuffles as binary digits, with the first shuffle equal to the most significant digit. Thus the shuffles In-In-Out-In produce the shift $1101_2 = 13_{10}$.

Example. Let's take a deck of 51 cards and look at what happens to the card in position 12 after four shuffles: In-In-Out-In. Knowing only the definitions of perfect shuffles, we can compute the movement of the card after each shuffle:

$$\mathbf{I}(12) \equiv 25 \ (\text{mod } 51);$$

$$\mathbf{I}(25) \equiv 51 \equiv 0 \ (\text{mod } 51);$$

$$\mathbf{O}(0) \equiv 0 \ (\text{mod } 51);$$

$$\mathbf{I}(0) \equiv 1 \ (\text{mod } 51).$$

Thus we have $\mathbf{I}\,\mathbf{O}\,\mathbf{I}\,\mathbf{I}(12) = 1$.

Using Theorem 2.6, we first compute the shift, $\sum_{i=1}^{4} 2^{k-i} \delta(S_i)$:

$$\delta(\mathbf{I})2^3 + \delta(\mathbf{I})2^2 + \delta(\mathbf{O})2^1 + \delta(\mathbf{I})2^0 = 1 \cdot 2^3 + 1 \cdot 2^2 + 0 \cdot 2^1 + 1 \cdot 2^0 = 13.$$

Now, following the theorem,

$$\mathbf{I}\,\mathbf{O}\,\mathbf{I}\,\mathbf{I}(12) \equiv 2^4 \cdot 12 + 13 \equiv 192 + 13 \equiv 205 \equiv 1 \ (\text{mod } 51).$$

Corollary 2.6. *In an odd deck of N cards, if $2^k \equiv 1 \ \text{mod}(N)$, then*

$$\mathbf{S}_k \cdots \mathbf{S}_2 \mathbf{S}_1(p) \equiv 2^k p + \sum_{i=1}^{k} 2^{k-i} \delta(\mathbf{S}_i) \ \text{mod}(N)$$

$$\equiv p + shift \, \text{mod}(N).$$

In other words, if $2^k \equiv 1 \ \text{mod}(N)$ the effect of *any* sequence of k perfect shuffles is to shift each card by a constant amount. This is equivalent to cutting *shift* cards from the bottom to the top of the deck. In a deck of 51 cards, any 8 perfect shuffles does this. Corollary 2.6 is used in the design of computer interconnection networks (see Chapter 5, Dynamic Computer Memories).

Example. This cutting by shuffling is best seen with a small example. Consider a deck of 15 cards, so $2^4 \equiv 1 \ (\text{mod } 15)$. We can shift 13 cards from the bottom to the top by doing the four shuffles In-In-Out-In, since $13_{10} = 1101_2$.

$$\mathbf{I}[0, 1, 2, 3, 4, 5, 6, 7, 8, 9, 10, 11, 12, 13, 14]$$
$$= [7, 0, 8, 1, 9, 2, 10, 3, 11, 4, 12, 5, 13, 6, 14]$$
$$\mathbf{I}[7, 0, 8, 1, 9, 2, 10, 3, 11, 4, 12, 5, 13, 6, 14]$$
$$= [3, 7, 11, 0, 4, 8, 12, 1, 5, 9, 13, 2, 6, 10, 14]$$
$$\mathbf{O}[3, 7, 11, 0, 4, 8, 12, 1, 5, 9, 13, 2, 6, 10, 14]$$
$$= [3, 5, 7, 9, 11, 13, 0, 2, 4, 6, 8, 10, 12, 14, 1]$$
$$\mathbf{I}[3, 5, 7, 9, 11, 13, 0, 2, 4, 6, 8, 10, 12, 14, 1]$$
$$= [2, 3, 4, 5, 6, 7, 8, 9, 10, 11, 12, 13, 14, 0, 1]$$

As one more example, consider $5_{10} = 0101_2$.

$$\mathbf{O}[0, 1, 2, 3, 4, 5, 6, 7, 8, 9, 10, 11, 12, 13, 14]$$
$$= [0, 8, 1, 9, 2, 10, 3, 11, 4, 12, 5, 13, 6, 14, 7]$$
$$\mathbf{I}[0, 8, 1, 9, 2, 10, 3, 11, 4, 12, 5, 13, 6, 14, 7]$$
$$= [11, 0, 4, 8, 12, 1, 5, 9, 13, 2, 6, 10, 14, 3, 7]$$
$$\mathbf{O}[11, 0, 4, 8, 12, 1, 5, 9, 13, 2, 6, 10, 14, 3, 7]$$
$$= [11, 13, 0, 2, 4, 6, 8, 10, 12, 14, 1, 3, 5, 7, 9]$$
$$\mathbf{I}[11, 13, 0, 2, 4, 6, 8, 10, 12, 14, 1, 3, 5, 7, 9]$$
$$= [10, 11, 12, 13, 14, 0, 1, 2, 3, 4, 5, 6, 7, 8, 9]$$

Corollary 2.7. *In an odd deck of N cards, if $p = 0$, then*

$$\mathbf{S}_k \cdots \mathbf{S}_2 \mathbf{S}_1(0) \equiv 2^k 0 + \sum_{i=1}^{k} 2^{k-i} \delta(\mathbf{S}_1) \mod(N)$$

$$\equiv shift \mod(N).$$

Moving a Card in a Deck

Corollary 2.7 gives magicians a clever way of moving the top card to any location in a deck. A version of Corollary 2.7 was first stated (without proof) by Alex Elmsley in his "Work in Progress." He provided a simple method of moving the top card to any location under the guise of shuffling the cards. (In his notation the top card is in position $p = 1$, while our notation has $p = 0$.) "To bring a card on top of the pack to any desired position, subtract one from the desired position, express the result as a binary number, and use it as instructions for a series of in- and out-shuffles, and the card will end where you want it." [**23**, 21]

Note that Elmsley's method works with odd or even decks because the top card never moves deeper than the middle (except, perhaps, with the last shuffle), and so no modular arithmetic is required to follow the top card. Using our notation of $p = 0$ for the top card, this can be expressed a little more clearly. To move the top card of a pack to any position, first express the position as a binary number. Then starting with the most significant digit, perform a perfect shuffle for each digit in the number—an Out for a 0 and an In for a 1.

Here's how to move the top card to position 14. First convert the final destination, 14, to a binary number: $14_{10} = 1110_2$. Then four perfect shuffles, In-In-In-Out, will move the top card, the "zeroth" card, to position 14. Murray Bonfeld showed how to move a card from the middle to the top of a deck [**9**]. S. Brent Morris extended Bonfeld's method to move any card to any position in odd decks [**73**], and Sarnath Ramnath and Daniel Scully developed a similar technique for even decks [**86**].

Corollary 2.7 and some statistics about American names fit together very nicely to make an entertaining card trick.

Trick 2.8. (**"A Spelling Bee"**) The magician tosses a deck of cards to anyone in the audience. The catcher chooses a card, signs her card with an "invisible" pen, reassembles the deck, and returns it. The magician shuffles the cards several times, ending with three perfect shuffles. The deck is tossed to someone else in the audience who tosses it again, and so on, until the final catcher helps find the chosen card. The catcher takes the deck from the case, and deals one card for each letter in his name. When the catcher finishes spelling his name, he has stopped at the card chosen and "signed" by the first catcher. (Tumultuous applause inevitably follows!)

Explanation. There are three distinct parts to this trick—the control, the placement, and the discovery—each equally important. The magician "controls" or keeps track of the chosen card by scratching its edge with his thumbnail. The chosen card is placed in position 4 with the three perfect shuffles ($4_{10} = 100_2 \rightarrow$ In-Out-Out). The card is put in position 4 because of a population parameter associated with American names: Most late twentieth century first names (perhaps

60–80%) have four or five letters. The deck is tossed around in the audience until caught by someone with four or five letters in their name.

Now for a more detailed explanation with performance notes.

The Control

I met a mathematician who said he wrote his doctoral dissertation on the mathematics of card shuffling. He claimed to be the only person in the world with a "Ph.D. in card shuffling." He taught me a trick while he showed me some different kinds of card shuffles. I'll show you the trick, but I'll need someone to help me. Here, catch the cards—and thanks for volunteering.

Toss the cards into the audience in a gentle, high arc. Ask the catcher's name (we'll assume it's Susan here), not so much to be friendly, but more to get the audience used to you asking for names. The name of the first catcher is unimportant, but the length of the name of the final catcher is crucial!

Thanks for helping me, Susan. Take the cards out of the case and shuffle them. I'll provide the play-by-play and color commentary while you mix the cards. When you finish, cut the deck into halves on the table in front of you.

Ask the catcher to shuffle the cards and to cut them into two approximately equal piles, cutting to the chosen card. The two piles should be left face down on the table in front of the catcher. If the bottom half of the deck with the chosen card is too small, it will be hard to mark the card with your thumbnail.

Susan, you shuffled the cards, and you chose this one by cutting to it. To make sure we can keep track of your card, I'm going to have you sign your name on the back with this pen. It's an official U.S. Government invisible pen, used to write the most secret and sensitive documents. Here, sign your name on the back of your card where there's a line in the pattern.

Introduce the "invisible" pen and have the chosen card "indelibly" marked. This looks like silly make-believe, but it has a critical purpose: it gives the magician a plausible reason for pointing to the back of the card where the "signature" goes. The action of pointing to the place on the card for the signature covers the thumbnail which is dragged along the edge of the top card. This control is

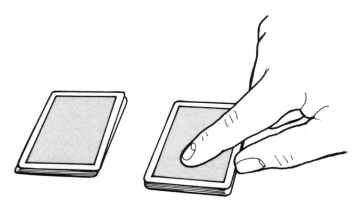

FIGURE 4.   Marking the edge of the top card with the thumbnail.

from "The Invisible Pen" [**84**], which gives more details and subtleties. (See figure 4.)

Now I can tell which card is yours because I have special lenses in my glasses (if you don't wear glasses say, "I'm wearing special contact lenses"). But to make it easier for you, why don't you look at the card and show it to everyone else. Now, put the deck back together and give it to me.

With the back of the card "signed," the catcher looks at and remembers the face, shows it to the rest of the audience, puts the deck back together, and returns it to the magician. Be sure someone besides the catcher sees the card— spectators often forget their card in the excitement of the trick. The location of the chosen card is unknown, but its scratched edge will be clearly visible on the side of the deck. The deck can be shuffled by the spectator several times before being returned.

The Placement

Let me get your card thoroughly lost in the deck. The mathematician told me there are several ways to shuffle cards—riffle, over-hand, Hindu, and so on—but he preferred the perfect shuffle. The perfect shuffle takes a little practice to master, but it does a superior job of mixing the cards.

The magician shuffles the cards several times, ending with three perfect shuffles. These different types of shuffles are explained in most standard card magic books. I illustrate the shuffles as I describe them, ever more thoroughly losing the chosen card (or so the audience should think). When I do a perfect shuffle I must stare intently at the side of the deck to estimate cutting it into exact halves. This gives me a perfect opportunity to locate the edge-scratched chosen card and to cut it to the top. Regardless of how few or how many cards I cut, I always say, "Oh darn, I didn't cut it exactly. Let me try again." This time I cut the deck exactly in half and do an in-shuffle, followed by two out-shuffles. Because $100_2 = 4_{10}$, the chosen card is now in location 4. (Remember that the top card is in location 0.)

The Discovery

Now to find your card, Susan, I'm going to need one more helper. I'll just toss the cards out again, which will give us another randomly selected volunteer.

The sleight-of-hand is over now, but the remaining showmanship is essential to a successfully entertaining trick. Toss the cards to someone else in the audience, and ask them their name; if they do not have four or five letters in their name, you'll have them toss the deck to someone else until you find someone with four or five letters in their name.

Let's make the selection of my volunteer even more random. Please toss the deck gently to someone else.

Continue until you find the right name. Empirical data indicates that 60–80% of "plain vanilla" American first names have four or five letters—Terry, Nancy, Mary, Brent, John, etc. I have never performed where there wasn't someone with the right number of letters in their name (well, hardly ever—see cases 3 and 4 below). Just continue having the deck tossed until the "right" person catches it.

Case 1. The final catcher has a four-lettered name.

Dave, thanks for volunteering. Take the deck out of the case and deal one card for each letter in your name. The next card in the deck, the one at which you stopped dealing, has Susan's "invisible signature" on the back. Please turn it over and show it to the audience.

Have the catcher deal one card for each letter in his name. (Cards 0, 1, 2, and 3 will be dealt.)

Case 2. The final catcher has a five-lettered name.

David, thanks for volunteering. Take the deck out of the case and deal one card for each letter in your name. The last card you dealt on the table, the one at which you stopped dealing, has Susan's "invisible signature" on the back. Please turn it over and show it to the audience.

Have the catcher deal one card for each letter in his name. (Cards 0, 1, 2, 3, and 4 will be dealt.)

Case 3. No one has a four- or five-lettered name in the first 3 or 4 tosses.

Elizabeth, thanks for volunteering. Take the deck out of the case and deal one card for each letter in your name. Put one of your hands on top of several cards. Let's take these cards away. You had free choices in selecting the cards, and only one remains. It has Susan's "invisible signature" on the back. Please turn it over and show it to the audience.

This doesn't happen often, but sometimes the deck will only be tossed to Bob, Ann, Jonathan, or Elizabeth. Or perhaps it's tossed to someone with a name that sounds like it should have 4 or 5 letters, but doesn't—Robyne or Dwayne, for example. More problems result from non-English names— Antoine, Siegfried, or Srinivas, for example. Once at the Citadel I asked a cadet his name and discovered they don't use their first names with instructors. He replied, "Cadet Williamson, Sir." When faced with one of these situations, I use an "out" that is effective, but not quite as dramatic as spelling exactly to the chosen card.

Make sure the catcher has more than five letters in their name (first and last, if necessary), and have one card dealt for each letter. Pay careful attention to where the chosen card is dealt! Say, "Put one of your hands on top of several cards." Note whether the chosen card is in the set covered or not. Then say, "Let's take away these cards." As you say this, casually remove the set of cards that does *not* contain the chosen card. There is now a reduced set of cards on the table, and the chosen card is contained among them.

This technique is called the "Magician's Choice," because there is no choice for the spectator. The magician never says in advance what will be done with the selected cards—if they will be removed or will be left on the table. For a "Magician's Choice" to succeed, the cards must be removed quickly and decisively, as if everything is going according to plan.

With the reduced set of cards, the spectator is again asked to cover or touch several cards. Remove the set that doesn't contain the chosen card. Never say in advance what will be done with the touched cards. When there are only 3 or 4 cards left, have the spectator use 2 or 3 fingers, touching one card with each finger. Ultimately the chosen card will remain on the table. Then you can say, "You had free choices in selecting the cards, and only one remains. It has the 'invisible signature' of the first catcher. Please turn it over and show it to the audience."

Case 4. Things really go wrong, and the final catcher unexpectedly has a three-lettered name.

This has happened only once, and it was certainly unnerving. I was speaking at the Goddard Space Flight Center and the deck was caught by "John." I asked him to spell his name, and he said, "Sure, J-O-N." I was ready for the grand climax, but the chosen card was not on top of the deck as I expected it to be, but in the second position.

I could have said, "Why don't you spell your last name too," and then continued as in Case 3. What I did was to say, "I foresaw that a 'John' would catch the cards, but I didn't pay attention to the spelling of his name. Would you please deal a card for the missing *h* and turn over the next card in the deck, *the one at which you should have stopped dealing.*" This was a flawed ending, but nonetheless puzzling. Fortunately this doesn't happen often!

Tumultuous applause spontaneously erupts accompanied by wild cheering. Well, this is the way I like to remember it.

3
Shuffle Groups

I was at a party once, and a fellow there brought out a deck of cards. He said he was a magician and asked someone to cut the cards. He perfectly shuffled the cards three times, with a few cuts in between. For good measure he cut the deck one last time.

The magician next dealt six five-card poker hands and invited anyone to play against his hand. He said his opponent could look at each of the other five hands and choose the one he liked the best before betting.

Well, I took him up. The first hand only had a pair of Jacks, the second was three of a kind, the third a flush, the fourth a straight, and the fifth four sevens. This was starting to look suspicious, but I decided to go with the four sevens.

With a sly wink the magician turned over his hand, a royal flush, and said, "You really didn't think I'd lose, did you?"

How did he do that?

There's an old story about the dairy department in the agriculture school of a large state university. They were trying to design an improved automatic milking machine. The equations describing the complex interactions between machine and cow were too difficult for them to solve, so in desperation they called in a professor from the mathematics department for help. After listening carefully to the problem, the mathematician left to work on the solution. A week or so later the professor returned with a thick paper. The dairy department asked expectantly, "Did you solve our problem?" To which the mathematician cheerfully replied, "I solved a similar but simplified version of the problem: Assume a spherical cow...."

Randomizing a Deck of Cards

The first significant mathematical result concerning the perfect shuffle came about when Solomon Golomb, a professor at the University of Southern California, tackled a tough, practical problem: How well do shuffles and cuts randomize a deck of playing cards? (The problem is at least practical to card players and casino owners.) Golomb's solution was not to consider a general "random" riffle shuffle as Bayer and Diaconis did thirty-one years later [7]. Rather, Golomb looked at the effect of perfect shuffles and cuts, a simplified version of the problem. His result is both neat and surprising [**33**].

First we define the Simple Cut, **C**, which cuts a single card from the top of a deck to the bottom. Any other cut of more cards can be obtained by repeating **C**. For example, cutting 10 cards from the top to the bottom is the same as **C**^{10}—repeating **C** 10 times.

Definition. The *Simple Cut*, **C**, on a deck of N cards cuts one card from the top to the bottom. The card in position p is moved by **C** as follows:

$$\mathbf{C}[0, 1, 2, \ldots, N - 1] = [1, 2, \ldots, N - 1, 0] \quad \text{or} \quad \mathbf{C}(p) \equiv p - 1 \bmod(N).$$

Definition. The *Terminal Transposition*, **T**, on a deck of N cards transposes the last two cards

$$\mathbf{T}[0, 1, 2, \ldots, N - 3, N - 2, N - 1] = [0, 1, 2, \ldots, N - 3, N - 1, N - 2].$$

Definition. The *Identity Transformation*, id, leaves the cards unmoved

$$\text{id}[0, 1, 2, \ldots, N - 2, N - 1] = [0, 1, 2, \ldots, N - 2, N - 1].$$

Definition. The *inverse of an operation*, denoted with the superscript -1, before or after an operation, results in the identity transformation

$$\mathbf{O}^{-1}[0, 1, 2, \ldots, N-2, N-1] = [0, 2, 4, \ldots, N-2, 1, 3, 5, \ldots, N-1]$$

$$\mathbf{OO}^{-1} = \mathbf{O}^{-1}\mathbf{O} = \text{id}$$

$$\mathbf{T}^{-1} = \mathbf{T}$$

$$\mathbf{T}^{-1}\mathbf{T} = \mathbf{TT}^{-1} = \mathbf{T}^2 = \text{id}.$$

Our next lemma shows that any permutation can be obtained from some combination of cuts and terminal transpositions.

Lemma 3.1. *The permutation group generated by the simple cut, $\mathbf{C}$, and the terminal transposition, $\mathbf{T}$, operating on a deck of N cards is the symmetric group, $\mathbf{S}_N$ of all $N!$ permutations.*

$$\langle \mathbf{C}, \mathbf{T} \rangle = \mathbf{S}_N.$$

Proof. Any permutation can be written as the product of two-cycles [35, 67]. We show that $\mathbf{C}$ and $\mathbf{T}$ generate any adjacent two-cycle, and that any two-cycle can be written as a product of adjacent two-cycles.

We use Δ to represent the deck in its original order. Consider the arbitrary adjacent two-cycle $(i, i+1)$.

$$\mathbf{C}^{i+2}[\Delta] = \mathbf{C}^{i+2}[0, 1, \ldots, i+1, i+2, \ldots, N-1]$$

$$= [i+2, \ldots, N-1, 0, 1, \ldots, i-1, i, i+1],$$

$$\mathbf{TC}^{i+2}[\Delta] = [i+2, \ldots, N-1, 0, 1, \ldots, i-1, i+1, i],$$

$$\mathbf{C}^{-(i+2)}\mathbf{TC}^{i+2}[\Delta] = [0, 1, \ldots, i-1, i+1, i, i+2, \ldots, N-1],$$

$$= (i, i+1).$$

We use the notation (i, j) to mean the elements in *positions* i and j are exchanged. The composition of cycles $(a, b)(c, d)$ is read from left to right, so first elements in positions a and b are exchanged, and then the elements in positions in c and d are exchanged. For an example with six elements,

$$(2, 3) = [0, 1, 3, 2, 4, 5]$$

and

$$(2, 3)(3, 4) = [0, 1, 3, 4, 2, 5].$$

Now consider the arbitrary two-cycle (i, j), $i < j$:

$$(i, i + 1) = [0, \ldots, i - 1, i + 1, i, i + 2, \ldots, N - 1],$$

$$(i, i + 1)(i + 1, i + 2)$$
$$= [0, \ldots, i - 1, i + 1, i + 2, i, i + 3, \ldots, N - 1],$$
$$(i, i + 1)(i + 1, i + 2) \cdots (j - 1, j)$$
$$= [0, \ldots, i - 1, i + 1, i + 2, \ldots, j - 1, j, i, j + 1, \ldots, N - 1],$$

$$(i, i + 1)(i + 1, i + 2) \cdots (j - 1, j)(j - 2, j - 1) \cdots (i, i + 1)$$
$$= (i, j). \qquad\qquad \square$$

Shuffles and Cuts in Even Decks

Golomb showed that the out-shuffle and simple cut on an even deck mix the deck very well indeed—every possible permutation can be generated.

Theorem 3.2. (Golomb) *In an even deck of* $N = 2n$ *cards, the out-shuffle,* **O**, *and the simple cut,* **C**, *generate the symmetric group* $\mathbf{S}_N$ *of all* $N!$ *permutations.*

Proof. Let $N = 2n$.

$$\mathbf{O}[0, 1, 2, \ldots, n - 1, n, \ldots, 2n - 2, 2n - 1]$$
$$= [0, n, 1, n + 1, \ldots, n - 2, 2n - 2, n - 1, 2n - 1],$$
$$\mathbf{C}^2\mathbf{O}[\Delta] = [1, n + 1, \ldots, n - 2, 2n - 2, n - 1, 2n - 1, 0, n],$$
$$\mathbf{TC}^2\mathbf{O}[\Delta] = [1, n + 1, \ldots, n - 2, 2n - 2, n - 1, 2n - 1, n, 0],$$
$$\mathbf{C}[\Delta] = [1, 2, 3, \ldots, n - 1, n, n + 1, \ldots, 2n - 2, 2n - 1, 0],$$
$$\mathbf{OC}[\Delta] = [1, n + 1, 2, n + 2, \ldots, n - 1, 2n - 1, n, 0].$$

We have $\mathbf{OC} = \mathbf{TC}^2\mathbf{O}$, and $\mathbf{T} = \mathbf{OCO}^{-1}\mathbf{C}^{-2}$. This means that the terminal transposition, **T**, is an element of the permutation group generated by **O** and **C**,

$$\mathbf{T} \in \langle \mathbf{O}, \mathbf{C} \rangle.$$

Thus

$$\langle \mathbf{T}, \mathbf{C} \rangle \subseteq \langle \mathbf{O}, \mathbf{C} \rangle,$$

but by Lemma 3.1, $\langle \mathbf{T}, \mathbf{C} \rangle = \mathbf{S}_N$, so $\langle \mathbf{O}, \mathbf{C} \rangle = \mathbf{S}_N$. $\quad\Box$

Theorem 3.3. *In an even deck of N cards, the in-shuffle,* **I**, *and the simple cut,* **C**, *generate the symmetric group* $\mathbf{S}_N$.

Proof. In an even deck, $\mathbf{IC} = \mathbf{TC}^2\mathbf{O}$, and the proof is the same as for Theorem 3.2. $\quad\Box$

Golomb's result on even decks, Theorem 3.2, provided some comfort to card players and casino owners: Cutting and out-shuffling can generate any permutation in a standard 52-card deck. However, we don't know what sequence of cuts and shuffles are required to generate a given permutation, nor do we know what happens when non-perfect shuffles are used. Nonetheless, our simplified model of shuffling and cutting doesn't put any theoretical limitation on the permutations we can generate nor any limitations on how well a deck of 52 cards theoretically can be mixed. The case of odd decks is surprisingly different.

Shuffles and Cuts in Odd Decks

Theorem 3.4. (Golomb) *In an odd deck of $N = 2n - 1$ cards, the out-shuffle,* **O**, *and the Simple Cut,* **C**, *generate the semi-direct product of the groups generated by* **O** *and* **C** *acting alone. That is,*

$$\langle \mathbf{O}, \mathbf{C} \rangle = \langle \mathbf{O} \rangle \times \langle \mathbf{C} \rangle,$$

and hence

$$|\langle \mathbf{O}, \mathbf{C} \rangle| = |\langle \mathbf{O} \rangle| \cdot |\langle \mathbf{C} \rangle| = o(\mathbf{O}, N)N.$$

Proof. Let $N = 2n - 1$.

$$\mathbf{O}[0, 1, 2, \ldots, n - 1, n, \ldots, 2n - 3, 2n - 2]$$
$$= [0, n, 1, n + 1, 2, \ldots, 2n - 3, n - 2, 2n - 2, n - 1],$$
$$\mathbf{C}^2\mathbf{O}[\Delta] = [1, n + 1, 2, \ldots, n - 2, 2n - 2, n - 1, 0, n],$$

$$C[\Delta] = [1, 2, 3, \ldots, n-1, n, n+1, \ldots, 2n-3, 2n-2, 0],$$

$$OC[\Delta] = [1, n+1, 2, \ldots, n-2, 2n-2, n-1, 0, n].$$

We have $OC = C^2O$, and any product of the form OC^a can be rewritten as $C^{2a}O$ by "moving" the C's to the left of the O. In particular, every permutation in $\langle O, C \rangle$ can be written as C^aO^b. Thus $|\langle O, C \rangle| \le N \cdot o(O, N)$.

Every power of the out-shuffle fixes the top card, while no power of C, except $C^N = $ id, fixes it. Hence the only powers in common of O and C are

$$O^{o(O,N)} = C^N = \text{id}.$$

If our inequality is strict, $|\langle O, C \rangle| < o(O, N)N$, then there exist a, b, c, and d such that

$$C^aO^b = C^cO^d$$

or

$$C^{a-c} = O^{d-b}.$$

But $C^{a-c} = O^{d-b}$ has only trivial solutions, so $a = c$, $b = d$, and $|\langle O, C \rangle| = o(O, N)N$.

Finally, we show that $\langle O, C \rangle$ is the semi-direct product of $\langle O \rangle$ and $\langle C \rangle$. To see that the subgroup $\langle C \rangle$ is normal in $\langle O, C \rangle$, $\langle C \rangle \lhd \langle O, C \rangle$, see that $AC^rA^{-1} \in \langle C \rangle$ for any r and all $A \in \langle O, C \rangle$. Recall that all elements of $\langle O, C \rangle$ can be written as C^aO^b and that $OC^a = C^{2a}O$. Consequently

$$O^bC^r = O^{b-1}OC^r$$

$$= O^{b-1}C^{2r}O$$

$$= O^{b-2}C^{4r}O^2$$

$$\vdots$$

$$= C^{2^br}O^b.$$

This means

$$AC^rA^{-1} = C^aO^bC^rO^{-b}C^{-a} = C^aC^{2^br}O^bO^{-b}C^{-a} = C^{2^br} \in \langle C \rangle,$$

so $\langle C \rangle \lhd \langle O, C \rangle$. We've shown above that $\langle C \rangle \cap \langle O \rangle = \{\text{id}\}$, and certainly $\langle O \rangle\langle C \rangle = \langle O, C \rangle$. Thus $\langle O, C \rangle$ is the semi-direct product of $\langle O \rangle$ and $\langle C \rangle$,

$$\langle O, C \rangle = \langle O \rangle \times \langle C \rangle. \qquad \square$$

Theorem 3.5. *In an odd deck of N cards,*

$$\langle \mathbf{I}, \mathbf{C} \rangle = \langle \mathbf{O}, \mathbf{I} \rangle = \langle \mathbf{O}, \mathbf{C} \rangle = \langle \mathbf{O} \rangle \times \langle \mathbf{C} \rangle.$$

Proof. Let $N = 2n - 1$.

$$\mathbf{I}[0, 1, 2, \ldots, n-2, n-1, \ldots, 2n-3, 2n-2]$$
$$= [n-1, 0, n, 1, n+1, 2, \ldots, 2n-3, n-2, 2n-2],$$
$$\mathbf{CI}[\Delta] = [0, n, 1, n+1, 2, \ldots, 2n-3, n-2, 2n-2, n-1],$$

but

$$\mathbf{O}[\Delta] = [0, n, 1, n+1, 2, \ldots, 2n-3, n-2, 2n-2, n-1].$$

We have $\mathbf{O} = \mathbf{CI}, \mathbf{O} \in \langle \mathbf{I}, \mathbf{C} \rangle$, and $\langle \mathbf{O}, \mathbf{C} \rangle \subseteq \langle \mathbf{I}, \mathbf{C} \rangle$. But in any product of $\mathbf{I}$'s and $\mathbf{C}$'s, $\mathbf{C}^{-1}\mathbf{O} = \mathbf{C}^{N-1}\mathbf{O}$ can be substituted for $\mathbf{I}$, so $\langle \mathbf{I}, \mathbf{C} \rangle \subseteq \langle \mathbf{O}, \mathbf{C} \rangle$. Thus $\langle \mathbf{I}, \mathbf{C} \rangle = \langle \mathbf{O}, \mathbf{C} \rangle$, and from Theorem 3.4, $\langle \mathbf{O} \rangle \times \langle \mathbf{C} \rangle = \langle \mathbf{O}, \mathbf{C} \rangle = \langle \mathbf{I}, \mathbf{C} \rangle$.

In the other case $\mathbf{OI}^{-1} = \mathbf{C}$, so $\mathbf{C} \in \langle \mathbf{O}, \mathbf{I} \rangle$, and in a similar fashion we can show $\langle \mathbf{O}, \mathbf{I} \rangle = \langle \mathbf{O}, \mathbf{C} \rangle = \langle \mathbf{O} \rangle \times \langle \mathbf{C} \rangle$. $\square$

Perfect shuffles (Out or In) and cuts on a deck of 52 cards generate $52! = 80,658,175,170,943,878,571,660,636,856,403,766,975,289,505,440,883,277,$ $824,000,000,000,000 \approx 80.7 \times 10^{66}$ permutations. By removing just one card to leave a deck of 51 cards, the perfect shuffles and cuts generate only $o(\mathbf{O}, 51) \times 51 = 8 \times 51 = 408$ permutations. We go from everything to almost nothing—a dramatic difference between odd and even decks!

When Golomb published his result there was some initial interest in using it to challenge the ESP research of Dr. J. B. Rhine at Duke University. Rhine used a deck of 25 cards, 5 each of five symbols: $\circ, \star, +, \approx$, and $\square$. Participants in the study tried to guess the order of symbols in a shuffled deck. If those conducting the study had used perfect shuffles and cuts to randomize the deck before each experiment, then only $o(\mathbf{O}, 25)25 = 20 \times 25 = 500$ permutations could have been generated. This compares to $25!/(5!)^5 = 623,360,743,125,120 \approx 6 \times 10^{14}$. As it turns out, Rhine and his researchers didn't use perfect shuffles, so this potential criticism never surfaced.

Out- and In-Shuffles in an Even Deck

Let us summarize our results so far with shuffle groups on a deck of N cards. If N is even, $\langle \mathbf{O}, \mathbf{C} \rangle = \langle \mathbf{I}, \mathbf{C} \rangle = S_N$, and if N is odd, $\langle \mathbf{O}, \mathbf{C} \rangle = \langle \mathbf{I}, \mathbf{C} \rangle =$

$\langle \mathbf{O}, \mathbf{I} \rangle = \langle \mathbf{O} \rangle \times \langle \mathbf{C} \rangle$. The only question left, for the sake of completeness, is to determine $\langle \mathbf{O}, \mathbf{I} \rangle$ for even decks. Determining $\langle \mathbf{O}, \mathbf{I} \rangle$, however, is much more difficult than meets the eye. (This is one of the fascinations of mathematics; "obvious" results can become virtually impossible with the slightest change. Witness the generalization of the Pythagorean theorem, $x^2 + y^2 = z^2$, which has an infinite number of integer solutions, (x, y, z), to Fermat's Last Theorem, $x^n + y^n = z^n$, which has no integer solutions for $n \geq 3$.)

It took twenty-two years from Golomb's paper in 1961 before $\langle \mathbf{O}, \mathbf{I} \rangle$ in even decks was nailed down. The result was due to Persi Diaconis, Ron Graham, and William Kantor [18]. The proof of their theorem is beyond the scope of this book, but it's too important not to state. Before stating their theorem, though, we need a preliminary consideration of the permutation group we're seeking to describe and some notation.

Recall "The Stay-Stack Principle" in chapter 1. In an even deck the cards in positions j and $N - 1 - j$—in "mirror" locations around the center of the deck—are moved to mirror locations by out- and in-shuffles. We can think of a deck of $2n$ cards as n pairs "locked" together in mirror positions. For example, rather than thinking of a deck of six cards as $[0, 1, 2, 3, 4, 5]$ when looking at the action of $\langle \mathbf{O}, \mathbf{I} \rangle$ we should think of the deck as $[3, 2, 1, -1, -2, -3]$. We can simplify the model further since $\langle \mathbf{O}, \mathbf{I} \rangle$ is a subgroup of permutations on $[\pm 3, \pm 2, \pm 1]$, if we agree that this is the top half of the deck, and matching each element in the top half is the element of opposite sign at the mirror location in the bottom half. This is a group of modified 3×3 permutation matrices, with entries $0, \pm 1$, and one nonzero entry in every row and column.

Another way to look at this group is to consider an octahedron centered around the origin in 3-space with six vertices $\pm e_1, \pm e_2, \pm e_3$ (see figure 1). Any symmetry of the octahedron that keeps vertices $\pm e_i$ opposite each other is a permutation in the group.

Definition. A *Weyl group* is a group W with a set of generators $s_1, s_2, \ldots, s_n$ such that $s_i^2 = \mathrm{id}$, and for the composition of each pair of generators $s_i s_j$, there is an integer n_{ij} such that $(s_i s_j)^{n_{ij}} = \mathrm{id}$, these being the only relations.

Definition. B_n is the Weyl group of $n \times n$ matrices with entries $0, \pm 1$, and one nonzero entry in every row and column, $|B_n| = n! \, 2^n$. Equivalently, B_n is the group of all $n! \, 2^n$ symmetries of the n-dimensional octahedron whose vertices are $\pm e_1, \ldots, \pm e_n$.

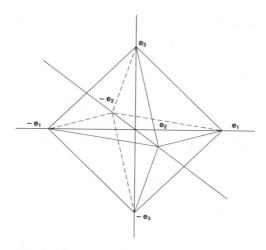

FIGURE 1. A 3-dimensional octahedron centered at the origin with vertices $\pm e_1, \pm e_2, \pm e_3$.

Definition. If $g \in B_n$, then sgn(g) is its sign as a permutation of $2n$ cards, and $\overline{\text{sgn}}(g)$ is its sign as a permutation of n centrally symmetric pairs.

Definition. The homomorphism $g \rightarrow \text{sgn}(g) \cdot \overline{\text{sgn}}(g)$ takes $B_n \rightarrow \{\pm 1\}$, and its kernel is the Weyl group D_n.

Definition. $PGL(2, 5)$ is the group of all linear fractional transformations over the finite field of five elements ($GF(5)$, the Galois Field of five elements).

Theorem 3.6. (Diaconis, Graham, and Kantor) *Let* $\langle \mathbf{O}, \mathbf{I} \rangle$ *be the permutation group generated by out- and in-shuffles on* $N = 2n$ *cards.*

- *If* $n \equiv 2 \bmod(4)$ *and* $n > 6$, $\langle \mathbf{O}, \mathbf{I} \rangle$ *is isomorphic to* B_n, *and* $|\langle \mathbf{O}, \mathbf{I} \rangle| = n! \, 2^n$.
- *If* $n = 6$, *that is,* $N = 12, \langle \mathbf{O}, \mathbf{I} \rangle$ *is the semi-direct product of* Z_2^n, *and* $PGL(2, 5)$ *and* $|\langle \mathbf{O}, \mathbf{I} \rangle| = n! \, 2^n / 3!$.
- *If* $n \equiv 1 \bmod(4)$ *and* $n \geq 5, \langle \mathbf{O}, \mathbf{I} \rangle$ *is the kernel of* $\overline{\text{sgn}}$ *and* $|\langle \mathbf{O}, \mathbf{I} \rangle| = n! \, 2^{n-1}$.
- *If* $n \equiv 3 \bmod(4)$, $\langle \mathbf{O}, \mathbf{I} \rangle$ *is isomorphic to* D_n, *and* $|\langle \mathbf{O}, \mathbf{I} \rangle| = n! \, 2^n - 1$.
- *If* $n \equiv 0 \bmod(4), n > 12$, *and* $n \neq 2^k, \langle \mathbf{O}, \mathbf{I} \rangle$ *is the intersection of* sgn *and* $\overline{\text{sgn}}$, *and* $|\langle \mathbf{O}, \mathbf{I} \rangle| = n! \, 2^{n-2}$.

- *If $n = 12$, that is, $N = 24$, $\langle \mathbf{O}, \mathbf{I} \rangle$ is the semi-direct product of Z_2^{11} and the Mathieu group M_{12} of degree 12 and $|\langle \mathbf{O}, \mathbf{I} \rangle| = n!\, 2^{n-1}/7!$.*
- *If $2n = 2^k$, $\langle \mathbf{O}, \mathbf{I} \rangle$ is the semi-direct product of k factors of Z_2, $Z_2 \times \cdots \times Z_2$, and Z_k, where Z_k acts by a cyclic shift; that is, the simple cut $\mathbf{C}$, and $|\langle \mathbf{O}, \mathbf{I} \rangle| = k2^n$.*

Table 2 gives values of $|\langle \mathbf{O}, \mathbf{I} \rangle|$ for $2n = 2 - 48$, to give a sense of the structure associated with the orders of the shuffle group in even decks.

TABLE 2   The Order of the Shuffle Group $\langle \mathbf{O}, \mathbf{I} \rangle$ on Even Decks

$2n$	$|\langle \mathbf{O}, \mathbf{I} \rangle|$	$2n$	$|\langle \mathbf{O}, \mathbf{I} \rangle|$
2	2	26	$M/2$
4	$2 \cdot 2^2$	28	M
6	$M/2$	30	$M/2$
8	$3 \cdot 2^2$	32	$5 \cdot 2^5$
10	$M/2$	34	$M/2$
12	$M/3!$	36	M
14	$M/2$	38	$M/2$
16	$4 \cdot 2^2$	40	$M/4$
18	$M/2$	42	$M/2$
20	M	44	M
22	$M/2$	46	$M/2$
24	$M/2 \cdot 7!$	48	$M/4$

$$(M = n!\, 2^n)$$

Theorem 3.5 has a simple corollary that is the basis of an amazing demonstration of poker playing "skills." It's worth mastering the perfect shuffle just for this trick.

Corollary 3.7. *In an odd deck of N cards, the product of a out-shuffles, b in-shuffles, and c cuts can be written as $\mathbf{C}^d \mathbf{O}^{a+b}$ for some d.*

Proof. From Theorem 3.5, $\mathbf{I} = \mathbf{C}^{-1}\mathbf{O} = \mathbf{C}^{N-1}\mathbf{O}$. From Theorem 3.4, $\mathbf{OC} = \mathbf{C}^2\mathbf{O}$, which shows how to move a $\mathbf{C}$ to the left of an $\mathbf{O}$ in the product $\mathbf{OC}$. Now substitute $\mathbf{C}^{N-1}\mathbf{O}$ for each $\mathbf{I}$ in the original product, and then move the $\mathbf{C}$'s to the left. $\qquad\square$

Trick 3.8. ("A Challenge Poker Deal") The magician gives a deck of cards several perfect shuffles, out or in, interspersed with any number of cuts. The cuts can be made by spectators. After mixing the cards, the magician deals several poker hands and challenges anyone to play against him. As the hands are examined, each one is found to be better than the one before. The magician, not surprisingly, has the winning hand.

Explanation. The mathematical results used in this trick are Corollary 3.7 and the fact that $o(\mathbf{O}, 51) = 8$. A deck of 51 cards is stacked in advance and out-shuffled $8 - 3 = 5$ times. Then three more shuffles returns the deck to its original stacked condition. Using an odd-sized deck allows cutting between the shuffles without disturbing the cyclic order of the cards.

This method of stacking a deck was first described in print in S. Victor Innis's 1915 *Inner Secrets of Crooked Card Players*. Innis determined that eight out-shuffles returned a deck of 52 to its original order. "By riffling a pack eight times with this riffle, each card is brought back to its original position." [**41**, 13] However, he didn't realize that five out-shuffles would put the deck three shuffles away from recycling. Rather he prescribed three tedious "inverse out-shuffles."

> Then I separate every other card from top to bottom, taking the top card to the left, the second to the right, and so on, placing the part of the pack on the left on top each time. I do this three times, then when the pack is cut in the middle and riffled every other card from the top to the bottom three times, this will bring the pack back to its original position ready to be dealt out. And the same hands will fall in their original respective places. [**41**, 14–15]

The Setup

1. *Use a deck of 51 cards and "stack" it for several poker hands.* I prefer to stack 6 hands, as more than this can become tedious. (Anyway, if the audience isn't impressed by you stacking 6 poker hands, they won't be impressed by 10!) I usually have the first hand very weak, perhaps a pair of jacks—just enough to open betting. Each hand is better than the one before, and I save a royal flush for myself.

2. *Make sure your "key" card is the bottom card in the stacked deck.* If you're borrowing someone's deck, put the joker on the bottom. If you're using your own deck, make a "corner short card" by carefully trimming a few millimeters from the back upper left and back lower right corners of one card [**37**, 518].

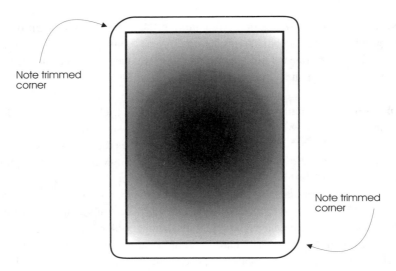

FIGURE 2.   Corner short card with upper left and lower right corners shortened.

3. *The final step before performance: Give the deck 5 perfect shuffles.* This sets up the deck for the trick. From Corollary 3.7, we know that for a deck of 51 cards, a total of 8 shuffles and any number of interspersed cuts has the same effect as $\mathbf{C}^d \mathbf{O}^8$, for some d. Since $o(\mathbf{O}, 51) = 8$, $\mathbf{C}^d \mathbf{O}^8 = \mathbf{C}^d \text{id} = \mathbf{C}^d$. In other words, for a deck of 51 cards, 8 perfect shuffles and any number of interspersed cuts leaves the deck in a cyclic shift of its original order. Because you gave the deck 5 shuffles before the performance, 3 more will return it to a cyclic shift of its original set up—stacked for the poker deal. (Three shuffles seems about right for a performance—any more and you lose your audience, any less and it doesn't seem quite enough to mix up the deck.)

The Performance

4. *Give the deck its 3 final shuffles interspersed with any number of cuts.* Always begin by openly cutting the cards, as in a card game. After the first perfect shuffle, immediately ask someone to cut the deck. Shuffle again, ask for another cut, and do one more shuffle. There's no need to ask for a third cut, as you will make this one, returning the deck to its original order.

You know that the deck is in its original order when your key card is on the bottom, and there are two ways to do this.

5a. *You're using a Joker as your key card.* Say, "Oh, I forgot to get rid of the jokers. This trick doesn't work with them in the deck." Run through the deck, cut the joker to the bottom, and discard it. (I don't know why this seems so reasonable to audiences—it must be because so many games can't be played with jokers.)

5b. *You're using a corner short card as your key card.* After the third perfect shuffle hold the deck in your left hand as if you're getting ready to deal. With your left thumb gently riffle down through the upper left corners of the cards. When you come to the corner short card, the riffling will almost stop on its own. (You may want to use a very shortened card initially until you get used to this.) Riffle again to this point in the deck and stop. The key card will be above your thumb. Cut this upper portion to the bottom of the deck.

6. *Deal the 6 poker hands.* I usually find that the audience isn't sure of what's going to happen, at least when they see the first few hands. However, as I go from 2 of a kind, to a 3 of a kind, to a full house, to a straight, to a flush, they usually realize that I've set things up. When I turn over my royal flush I wink and say, "You really didn't think I'd lose, did you?"

4
Generalizing the Perfect Shuffle

I saw a guy do a very strange card trick. He cut the deck into three parts and shuffled the parts together, without finishing the shuffle. It sure looked awkward for him to hold.

He had two spectators each choose a card, and then he chose one himself. Each card came from a different part of the deck. With three chosen cards—two by spectators and one by the magician—the deck was reassembled. The magician cut and shuffled the deck several times. He then cut the deck into three parts and shuffled the three parts into each other in a way I didn't think possible.

When this triple shuffle was done (which was almost worth applauding itself), he spread out the cards. There in the middle of the deck was his card, still face up. He then turned over the two cards on either side of his, and they were the other two chosen cards. Eyes bulged and jaws dropped.

How did he do that?

Someone once suggested the following sign should be prominently posted in the coffee lounge of every mathematics graduate department:

BE WISE; GENERALIZE!

Certainly generalization is one of the first research techniques learned in mathematics. Students are subjected regularly to "twiddling" with hypotheses and to seeing what happens when conditions are loosened or eliminated. The most-admired generalizations are those that in some sense are "natural," for example Fermat's generalization of the Pythagorean Theorem from $x^2 + y^2 = z^2$, with an infinite number of integer solutions, to $x^n + y^n = z^n$, with no integer solutions. (My non-mathematical friends will argue there's nothing at all natural about anything in mathematics, but that's another story.)

Out-Shuffling Several Packets of Cards

The out-shuffle leads to a natural generalization when one thinks about what happens to the cards: The deck is divided into two halves and the halves are interlaced. You can visualize the two halves being riffled together, as in figure 1, with cards falling precisely from one half and then another.

FIGURE 1.   The out-shuffle: two halves interlaced with the top card left on top.

Imagine this generalization of an out-shuffle: The deck of cards is cut into thirds and each third is held ready to be riffled. Visualize the three thirds being riffled together, cards falling precisely from the bottom third, the middle third, the top third, and then starting over. This is a generalization of the out-shuffle to three hands (hard to perform, but easy to imagine—see figure 2).

You can also imagine a deck of cards cut into fourths and cards falling from each fourth in order from the bottom fourth, next-to-bottom, next-to-top,

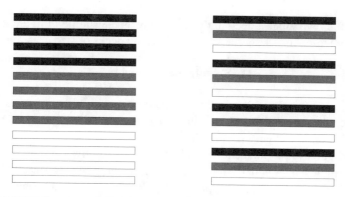

FIGURE 2. A generalized out-shuffle: three thirds of a deck interlaced.

and top, as in figure 3. We have yet another generalization of the perfect shuffle, this time to four hands.

In a similar way, the generalizations can start algebraically from the equation defining the ordinary out-shuffle. If the deck of cards is even, $N = 2n$, then

$$\mathbf{O}(p) \equiv 2p \bmod (2n - 1).$$

Generalizing everything just one step, we define a "generalized out-shuffle of order 3," or a "3-shuffle" on a deck of $N = 3n$ cards as

$$\mathbf{O}_3(p) \equiv 3p \bmod (3n - 1).$$

FIGURE 3. A generalized out-shuffle: four fourths of a deck interlaced.

This shuffle rearranges a deck in the same way as the shuffle in figure 2. The shuffle in figure 3 is a 4-shuffle, and is defined on decks of size $N = 4n$:

$$\mathbf{O}_4(p) \equiv 4p \bmod (4n - 1).$$

The physical generalization and the algebraic one produce the same result, at least when the size of the deck is a multiple of the number of packets shuffled together, or one less than such a multiple.

Karl Fulves first pursued the triple shuffle—as a physical, not algebraic generalization, but without the realization it can actually be performed [**27**, **45**]. (For performance details, see Appendix 2, "How to Do the Faro Shuffle: The Triple Faro Shuffle.") S. Brent Morris independently generalized the faro shuffle and published the full algebraic generalization in his doctoral dissertation at Duke University [**76**].

Definition. The generalized perfect out-shuffle of order k, or k-shuffle, $\mathbf{O}_k$, is defined on a deck of $N = kn$ or $N = kn - 1$ cards as $\mathbf{O}_k(p) \equiv kp \bmod (kn - 1)$, $0 \le p < kn - 2$, and $\mathbf{O}_k(kn - 1) = kn - 1$.

Just as with the ordinary out-shuffle, or 2-shuffle, the order of a k-shuffle for a deck of $N = kn$ cards, $o(\mathbf{O}_k, N)$, is the smallest number of shuffles that returns the deck to its original order. The extension carries over naturally from 2-shuffles: $o(\mathbf{O}_k, N)$ is the order of $k \bmod (N - 1)$. Thus $k^{o(\mathbf{O}, N)} \equiv 1 \bmod (N - 1)$, and $|\langle \mathbf{O}_k \rangle| = o(\mathbf{O}_k, N)$.

We can continue the physical generalization "naturally" when the deck is not a multiple of k, but we need a slightly different model of shuffling. Consider first a different method of performing the 2-shuffle on a deck of $N = 2n$ cards. Lay out the deck in N positions and pick up card 0 as the new top card. Move to card $0 + n$ and pick it up as the next-to-the-top card. Now pick card 1, then card $1 + n$, and so on. After N steps the reassembled deck has been rearranged by a 2-shuffle, $\mathbf{O}_2$. If there are $N = 2n + 1$ cards in the deck, the procedure is almost the same. The cards are laid out as before, but they are picked up in order $0, n + 1, 1, n + 2, 2, \ldots, n - 1, 2n, n$.

Imagine now doing a 3-shuffle, $\mathbf{O}_3$, on a deck of size $N = 3n + 1$. Divide the deck into 3 parts and let the first part have $n + 1$ cards and the second and third parts n cards each. Then the cards are picked up in order $0, n + 1, 2n + 1$, $1, n + 2, 2n + 2, \ldots, n - 1, 2n, 3n, n$. This is a "natural" way to extend the idea of a generalized perfect out-shuffle to a deck of any size.

Another way to look at the generalized shuffle $\mathbf{O}_k$ is as a card game with k players. (One really can't have too many ways of looking at a math problem.) If the deck has $N = nk + q$ cards, $0 \le q < k$, the first q players each get, in turn, $n + 1$ cards from the top of the deck and the next $k - q$ players get n cards each. The first card of the first player becomes the top card of the deck, the first card of the second player becomes the next-to-the-top card, and so on. After n "plays," the last $k - q$ players exhaust their cards, and the $n + 1$st play is limited to the first q players, with the last card of the qth player becoming the bottom card of the deck.

Looking for a Neat Formula

These physical generalizations work nicely and help us visualize what a generalized out-shuffle should do. The algebraic generalization to decks of size $N = nk + q$, however, is not quite so neat. If we keep the formula for $\mathbf{O}_k(p)$ fixed, then the modulus must change for different values of p.

$$\mathbf{O}_k(p) \equiv kp \begin{cases} \operatorname{mod}\big((n + 1)k - 1\big) & 0 \le p < q(n + 1) + n \\ \operatorname{mod}\big((q + 1)(nk - 1) + nq\big) & q(n + 1) + n \le p \\ & \qquad < q(n + 1) + 2n \\ \operatorname{mod}\big((q + 2)(nk - 1) + nq\big) & q(n + 1) + 2n \le p \\ & \qquad < q(n + 1) + 3n \\ \qquad\qquad \vdots & \qquad\qquad \vdots \\ \operatorname{mod}\big((q + r)(nk - 1) + nq\big) & q(n + 1) + rn \le p \\ & \qquad < q(n + 1) + (r + 1)n \\ \qquad\qquad \vdots & \qquad\qquad \vdots \\ \operatorname{mod}\big((k - 1)(nk - 1) + nq\big) & q(n + 1) + (k - q - 1)n \le p \\ & \qquad < q(n + 1) + (k - q)n \\ & \qquad = kn + q \end{cases}$$

Here's one way to think of the process. The first q players, each with $n + 1$ cards, as well as the $q + 1$st player, the first with n cards, use the same modulus as for $\mathbf{O}_k$ on a deck of $(n + 1)k$ cards, $(n + 1)k - 1$. [Their cards move to the same positions as if the deck was of size $(n + 1)k$.] Each of the next $k - q - 1$ players, each with n cards, uses a different modulus. For the case of $\mathbf{O}_5$ for $N = 17 = 3 \cdot 5 + 2$ (shown in figure 4), the next equation gives the modular relationships describing $\mathbf{O}_5$, with $k = 5$, $n = 3$, and $q = 2$.

FIGURE 4. O_5 on a deck of size $N = 17 = 3 \cdot 5 + 2$.

$$\mathbf{O}_5(p) \equiv 5p \begin{cases} \mathrm{mod} \left((3+1)5 - 1 = 19 \right) & 0 \le p < 11 \\ \mathrm{mod} \left((3+1)(2 \cdot 5 - 1) + 2 \cdot 5 = 46 \right) & 11 \le p < 14 \\ \mathrm{mod} \left((3+2)(2 \cdot 5 - 1) + 2 \cdot 5 = 55 \right) & 14 \le p < 17 \end{cases}$$

Instead of keeping the formula for $\mathbf{O}_k$ fixed, we can keep the modulus fixed at $(n+1)k - 1$. This arrangement isn't much neater than when we kept the formula fixed.

$$\mathbf{O}_k(p) \equiv \left. \begin{cases} kp & 0 \le p < q(n+1) + n \\ k(p+1) & q(n+1) + n \\ & \quad \le p < q(n+1) + 2n \\ k(p+2) & q(n+1) + 2n \\ & \quad \le p < q(n+1) + 3n \\ \vdots & \vdots \\ k(p+r) & q(n+1) + rn \\ & \quad \le p < q(n+1) + (r+1)n \\ \vdots & \vdots \\ k(p+k-q-1) & q(n+1) + (k-q-1)n \le p \\ & \quad < q(n+1) + (k-q)n \\ & \quad = kn + q \end{cases} \right\} \mathrm{mod} \left((n+1)k - 1 \right)$$

For a deck of $N = 17 = 3 \cdot 5 + 2$, the following equation shows the formulas for $\mathbf{O}_5$.

$$\mathbf{O}_5(p) \equiv \begin{cases} 5p & 0 \le p < 11 \\ 5(p+1) & 11 \le p < 14 \\ 5(p+2) & 14 \le p < 17 \end{cases} \bmod \big((3+1)5 - 1 = 14\big)$$

We can take the "best of both worlds" and get a compact expression for $\mathbf{O}_k$ by using two equations and moduli. One equation and modulus is used for the first $q(n+1) + n$ cards (q players with $n+1$ cards each plus the first player with n cards), and then another for the last $(k-q)n - n$ cards. In fact, the next equation really gives us a pretty good expression for $\mathbf{O}_k$—compact, simple, and easy to use.

$$\mathbf{O}_k(p) \equiv \begin{cases} kp \bmod \big((n+1)k - 1\big) & 0 \le p < q(n+1) + n \\ k(p-q) \bmod (nk - 1) & q(n+1) + n \le p < nk + q \end{cases}$$

Note that when $q = 0$, $N = nk$, and the above equation reduces to $\mathbf{O}_k(p) \equiv kp(\bmod\, nk - 1)$; when $q = k - 1$, $N = (n+1)k - 1$, and the equation reduces to $\mathbf{O}_k(p) \equiv kp(\bmod(n+1)k - 1)$. This is exactly the situation for the ordinary out-shuffle, $\mathbf{O}_2$.

Permutation Matrices

The equation just discussed is convenient to work with if you want to know where cards move, but it's not a good representation for group work. An approach that works well for groups is to look at the permutation matrix $P_{k,N} \in \mathbb{R}^{N \times N}$ that corresponds to $\mathbf{O}_k$ on a deck of size $N = nk + q$.

$$P_{k,N} \begin{bmatrix} 0 \\ 1 \\ \vdots \\ N-1 \end{bmatrix} = \big(\mathbf{O}_k[0, 1, \ldots, N-1]\big)^{\mathrm{T}}$$

The matrix $P_{k,N}$ has the following block-structure, which comes directly from the card game interpretation of $\mathbf{O}_k$ on $N = nk + q$ cards:

$$P_{k,N} = \begin{matrix} nk \left\{ \vphantom{\begin{matrix}E\\ \vdots \\ E\end{matrix}} \right. \\ q \left\{ \vphantom{G} \right. \end{matrix} \begin{bmatrix} E_{11} & E_{21} & \cdots & E_{q1} & F_{q+1,1} & \cdots & F_{k1} \\ \vdots & \vdots & & \vdots & \vdots & & \vdots \\ E_{1n} & E_{2n} & \cdots & E_{qn} & F_{q+1,n} & \cdots & F_{kn} \\ G_{1,n+1} & G_{2,n+1} & \cdots & G_{q,n+1} & 0 & \cdots & 0 \end{bmatrix},$$

where $E_{ij} \in \mathbb{R}_{k \times (n+1)}$, $F_{ij} \in \mathbb{R}_{k \times n}$, and $G_{ij} \in \mathbb{R}_{q \times (n+1)}$ are all unit matrices with 1 in the (i, j) position and zero elsewhere. The k block-columns of $P_{k,N}$ correspond to the k players, while the $n + 1$ block-rows correspond to the $n + 1$ "rounds" of the card game.

Interpreting $P_{k,N}$ in this light, E_{11} corresponds to the first card of the shuffled deck coming from the first player (who holds $n + 1$ cards), E_{21} is the second card of the shuffled deck coming from the second player (who also holds $n + 1$ cards), $F_{q+1,1}$ is the $q + 1$st card coming from the $q + 1$st player (who holds n cards), and so on. Finally when we get to the G_{ij}'s in $P_{k,N}$, only q cards are left in the deck, each belonging to one of the first q players.

The simple cut $\mathbf{C}$ on a deck of size N (sometimes written $\mathbf{C}_N$ to emphasize the size of the deck) has an uncomplicated representation as a permutation matrix. If $\mathbf{I}_m$ is the $m \times m$ identity matrix, then

$$\mathbf{C}_N = \mathbf{C} = \begin{bmatrix} 0 & 1 & \cdots & 0 \\ \vdots & & \ddots & \\ 0 & 0 & \cdots & 1 \\ 1 & 0 & \cdots & 0 \end{bmatrix}, \quad \text{and} \quad \mathbf{C}^k = \begin{bmatrix} 0 & \mathbf{I}_{N-k} \\ \mathbf{I}_k & 0 \end{bmatrix}.$$

Recall that in calculating the group structure generated by $\mathbf{O}_2$ and $\mathbf{C}$, the terminal transposition, $\mathbf{T}$ (a transposition or cyclic shift of the last 2 cards) is important. The natural generalization of $\mathbf{T}$ is a cyclic shift of the last k cards, which we denote by $\mathbf{C}^{(k)}$.

$$\mathbf{C}^{(k)}[0, \ldots, N - k - 1, N - k, N - k + 1, \ldots, N - 1]$$

$$= [0, \ldots, N - k - 1, N - k + 1, \ldots, N - 1, N - k].$$

The matrix corresponding to $\mathbf{C}^{(k)}$ is

$$\mathbf{C}^{(k)} = \begin{bmatrix} \mathbf{I}_{N-k} & 0 \\ 0 & \mathbf{C}_k \end{bmatrix}, \quad \text{so that} \quad \left(\mathbf{C}^{(k)}\right)^r = \begin{bmatrix} \mathbf{I}_{N-k} & 0 \\ 0 & \mathbf{C}_k^r \end{bmatrix}.$$

Generalizing Theorems

We now have the notation to present a generalization of the relationship $\mathbf{OC} = \mathbf{TC}^2\mathbf{O}$ from Theorem 3.2. The relationship covers any k-shuffle and decks of all sizes; it confirms that our pursuit of generalizations is following a "natural" path. The result was first proved by S. Brent Morris in his 1974 dissertation at Duke University [76] by a very tedious analysis of the permutations involved.

Morris and Robert E. Hartwig came up with a more elegant proof [74] using permutation matrices presented here.

Theorem 4.1. (Morris & Hartwig) *In a deck of* $N = nk + q$ *cards*

$$\mathbf{O}_k \mathbf{C} = \left(\mathbf{C}^{(k)}\right)^{q+1} \mathbf{C}^k \mathbf{O}_k.$$

Proof. Consider the product, $\mathbf{C}^k P_{k,N}$ and its corresponding permutation matrix

$$
\mathbf{C}^k P_{k,N} = \begin{bmatrix}
E_{12} & \cdots & E_{q2} & F_{q+1,2} & \cdots & F_{k2} \\
\vdots & & \vdots & \vdots & & \vdots \\
E_{1n} & \cdots & E_{qn} & F_{q+1,n} & \cdots & F_{kn} \\
G_{1,n+1} & \cdots & G_{q,n+1} & 0 & \cdots & 0 \\
E_{11} & \cdots & E_{q1} & F_{q+1,1} & \cdots & F_{k1}
\end{bmatrix}
$$

Multiplying each side by $\left(\mathbf{C}^{(k)}\right)^{q+1}$ yields

$$
\left(\mathbf{C}^{(k)}\right)^{q+1} \mathbf{C}^k P_{k,N} = \begin{bmatrix}
E_{12} & \cdots & E_{q2} & F_{q+1,2} & \cdots & F_{k2} \\
\vdots & & \vdots & \vdots & & \vdots \\
E_{1n} & \cdots & E_{qn} & F_{q+1,n} & \cdots & F_{kn} \\
G_{1,n+1} & \cdots & G_{q,n+1} & 0 & \cdots & 0 \\
\mathbf{C}_k^{q+1} E_{11} & \cdots & \mathbf{C}_k^{q+1} E_{q1} & \mathbf{C}_k^{q+1} F_{q+1,1} & \cdots & \mathbf{C}_k^{q+1} F_{k1}
\end{bmatrix},
$$

where the last block-row simplifies to

$$[E_{k-q,1}, E_{k-q-1,1}, \ldots, E_{k-1,1}, F_{k1}, F_{11}, F_{22}, \ldots, F_{k-q-1,1}],$$

on using the identities $\mathbf{C}_k^{q+1} E_{i1} = E_{k-q+i-1,1}$, and $\mathbf{C}_k^{q+1} F_{i1} = F_{k-q+i-1,1}$. Note that premultiplication by $\mathbf{C}_k$ has the effect of moving the top row to the bottom. Thus $\mathbf{C}_k E_{i1} = E_{i-1,1}$ for $i \neq 1$, and $\mathbf{C}_k E_{11} = E_{k1}$.

On the other hand, $P_{k,N} \mathbf{C}$ is obtained from $P_{k,N}$ by bringing the last column of $P_{k,N}$ to the front and repartitioning. This process replaces E_{ij} by $E_{i,j+1}$ and F_{ij} by $F_{i,j+1}$ in the matrix $P_{k,N}$. Hence the first $n - 1$ block-rows of blocks in $P_{k,N} \mathbf{C}$ and the preceeding equation agree, and we are only left with the proof of the identity

$$
\begin{matrix} k\{ \\ q\{ \end{matrix} \begin{bmatrix}
E_{1n} & \cdots & E_{qn} & F_{q+1,n} & \cdots & F_{km} \\
G_{1,n+1} & \cdots & G_{q,n+1} & 0 & \cdots & 0
\end{bmatrix} \mathbf{C}
$$

$$
= \begin{matrix} q\{ \\ k\{ \end{matrix} \begin{bmatrix}
G_{1,n+1} & \cdots & G_{q,n+1} & 0 & \cdots & 0 \\
E_{k-q,1} & \cdots & E_{k-1,1} & F_{k1} & \cdots & F_{k-q-1,1}
\end{bmatrix}.
$$

On performing the simple cut and partitioning off the first q rows, it follows directly that both sides are equal. $\square$

Note that Theorem 4.1 contains the main results of Theorems 3.2 and 3.4 for $\mathbf{O}_2$ and $\mathbf{C}$. When $k = 2$ and $q = 0$, N is even, and $\mathbf{O}_2\mathbf{C} = (\mathbf{C}^{(2)})^{0+1}\mathbf{C}^2\mathbf{O}_2 = \mathbf{TC}^2\mathbf{O}_2$, and when $k = 2$ and $q = 1$, N is odd, and $\mathbf{O}_2\mathbf{C} = (\mathbf{C}^{(2)})^{1+1}\mathbf{C}^2\mathbf{O}_2 = \mathbf{C}^2\mathbf{O}_2$.

Here's a quick example of Theorem 4.1 for $N = 3 \cdot 4 + 2 = 14$, $n = 3, k = 4, q = 2$.

$$\Delta = [0, 1, 2, 3, 4, 5, 6, 7, 8, 9, 10, 11, 12, 13]$$

$$\mathbf{O}_4(\Delta) = [0, 4, 8, 11, 1, 5, 9, 12, 2, 6, 10, 13, 3, 7]$$

$$\mathbf{O}_4\mathbf{C}(\Delta) = [1, 5, 9, 12, 2, 6, 10, 13, 3, 7, \mathbf{11}, \mathbf{0}, \mathbf{4}, \mathbf{8}]$$

$$\mathbf{C}^4\mathbf{O}_4(\Delta) = [1, 5, 9, 12, 2, 6, 10, 13, 3, 7, \mathbf{0}, \mathbf{4}, \mathbf{8}, \mathbf{11}]$$

$$(\mathbf{C}^{(4)})^3\mathbf{C}^4\mathbf{O}_4(\Delta) = [1, 5, 9, 12, 2, 6, 10, 13, 3, 7, \mathbf{11}, \mathbf{0}, \mathbf{4}, \mathbf{8}]$$

Corollary 4.2. (Morris & Hartwig) *In a deck of $N = nk + q$ cards, if $\mathbf{C}^{[k]}$ is the simple cut on the first k cards, then*

$$\mathbf{O}_k\mathbf{C} = \mathbf{C}^k \left(\mathbf{C}^{[k]}\right)^{q+1} \mathbf{O}_k.$$

We have extended the definition of the out-shuffle from physical and algebraic models, and both extensions agree. (Score one for the naturalness of the extensions.) The generalizations nicely carry over the key relationships between $\mathbf{O}_2$ and $\mathbf{C}$, $\mathbf{O}_2\mathbf{C} = \mathbf{TC}^2\mathbf{O}_2$ in even decks, and $\mathbf{O}_2\mathbf{C} = \mathbf{C}^2\mathbf{O}_2$ in odd decks. This work leads to the group generated by $\mathbf{O}_k$ and $\mathbf{C}$, $\langle\mathbf{O}_k, \mathbf{C}\rangle$. Before we reach that goal, however, we need a few preliminary definitions and results.

Definition. If G is a permutation group on the set of elements Ω, and $\Delta_i \subset \Omega$ is a subset of Ω, then $\{\Delta_1, \Delta_2, \ldots, \Delta_t\}$ is a *system of imprimitivity* for G provided the following hold:

(1) $\Omega = \Delta_1 \cup \Delta_2 \cup \cdots \cup \Delta_t$;

(2) $\Delta_i \cap \Delta_j = \emptyset$ for $i \neq j$;

(3) For each i there exists a j such that $g(\Delta_i) = \Delta_j, g \in G$.

In a system of imprimitivity, the Δ_i are called *sets of imprimitivity* or *blocks*. Condition (3) says that each permutation either maps a set of imprimitivity onto itself, or onto another set of imprimitivity. It also says $|\Delta_i| = |\Delta_j|, \forall\, i, j$.

Definition. If G is a permutation group on the set of elements $\Omega = \{\omega_i\}$, and the only systems of imprimitivity for G are trivial, that is, $\{\Omega\}$ or $\{\{\omega_1\}, \{\omega_2\}, \ldots, \{\omega_r\}\}$, then G is a *primitive permutation group*.

Lemma 4.3. (Morris & Hartwig) *In a deck of* $N = nk + q$ *cards*, $0 \le q < k$, *if* $d = \gcd(q + 1, k)$, *the group* $\langle \mathbf{C}, (\mathbf{C}^{(k)})^d \rangle$ *generated by the simple cut* $\mathbf{C}$ *on* N *elements and the dth power of the simple cut* $\mathbf{C}^{(k)}$ *on the last k elements is:*

(1) if $N = k$ *or if* $q = k - 1$, *the cyclic group of order N generated by* $\mathbf{C}$;

(2) if N is odd, the alternating group A_N;

(3) if N is even, the symmetric group S_N.

Proof. If $N = k$ (and thus $n = 1$ and $q = 0$), then $\gcd(q + 1, k) = 1, \mathbf{C}^{(k)} = \mathbf{C}$, and $\langle \mathbf{C}, (\mathbf{C}^{(k)})^d \rangle = \langle \mathbf{C}, \mathbf{C}^1 \rangle = \langle \mathbf{C} \rangle$. If $q = k - 1$, then $\gcd(q + 1, k) = k = d$, and $\langle \mathbf{C}, (\mathbf{C}^{(k)})^d \rangle = \langle \mathbf{C}, (\mathbf{C}^{(k)})^k \rangle = \langle \mathbf{C}, \mathrm{id} \rangle = \langle \mathbf{C} \rangle$.

Suppose now that $N \ne k$ and $q \ne k - 1$, the latter of which implies that $\gcd(q + 1, k) \le \frac{1}{2}k$. We will show that $\langle \mathbf{C}, (\mathbf{C}^{(k)})^d \rangle$ is a primitive permutation group. First, any set of imprimitivity of $\langle \mathbf{C}, (\mathbf{C}^{(k)})^d \rangle$ is a set of imprimitivity of $\langle \mathbf{C} \rangle$ and must be of the form

$$A_{i,s} = \{i + js \mid 1 \le i \le s; s \mid N; 0 \le j \le (N/s) - 1\},$$

which is a residue class for s, a divisor of N. That is because all elements of $A_{i,s}$ are shifted by elements of $\langle \mathbf{C} \rangle$ either onto themselves or onto another residue class for s. Thus, if one residue class of s, $A_{i,s}$, is a set of imprimitivity of $\langle \mathbf{C} \rangle$, then $A_{j,s}$ is also, for all $j, 1 \le j \le s$.

Assuming $A_{i,s}$ is a set of imprimitivity for $\langle \mathbf{C}, (\mathbf{C}^{(k)})^d \rangle$ with $s \le \frac{1}{2}k$, choose $x, y \in A_{i,s}$ such that $y < N - k \le x + d$. Since $\mathbf{C}^{(k)}$ only moves the last k cards, $(\mathbf{C}^{(k)})^d(y) = y$, and because $A_{i,s}$ is a set of imprimitivity, $(\mathbf{C}^{(k)})^d(A_{i,s}) = A_{i,s}$. Since $(\mathbf{C}^{(k)})^d$ shifts each of the last k cards by $d, (\mathbf{C}^{(k)})^d(x + d) = x$, and we have $x + d \in A_{i,s}$. Since $x = i + js$, and $x + d \in A_{i,s}$, then $s \mid d$. As $s \mid N$ and $(d, N) = 1$, we have $s = 1$, and hence there are no nontrivial sets of imprimitivity for $\mathbf{C}$. Thus $\langle \mathbf{C} \rangle$ and $\langle \mathbf{C}, (\mathbf{C}^{(k)})^d \rangle$ are primitive permutation groups.

If G is a primitive permutation group and contains a 3-cycle, then G is either the alternating or symmetric group [**101**, 34]. We will show that $\mathbf{C}$ and $(\mathbf{C}^{(k)})^d$ generate a 3-cycle; then all that remains is to determine if the generators $\mathbf{C}$ and $(\mathbf{C}^{(k)})^d$ are even or odd permutations.

For $d = 1$, we can generate the 3-cycle $(0, k, 1)$ by

$$\mathbf{C}^{-1}(\mathbf{C}^{(k)})^{-1}\mathbf{C}\mathbf{C}^{(k)}[0, 1, 2, \ldots, k - 1, k, k + 1, \ldots]$$
$$= [1, k, 2, \ldots, k - 1, 0, k + 1, \ldots],$$

while for $d > 1$, we can generate the 3-cycle $(0, k, d)$ by using $(\mathbf{C}^{(k)})^d$ for $\mathbf{C}^{(k)}$.

If k is odd, $\mathbf{C}^{(k)}$ is an even permutation, and $(\mathbf{C}^{(k)})^d$ is even for all d. If k is even, $\mathbf{C}^{(k)}$ is an odd permutation, but $k + 1$ and $d = \gcd(k + 1, q)$ are odd, so $(\mathbf{C}^{(k)})^d$ is an even permutation. Regardless of the value of k, $(\mathbf{C}^{(k)})^d$ is always an even permutation and thus does not affect whether $\langle \mathbf{C}, (\mathbf{C}^{(k)})^d \rangle$ is A_N or S_N. If N is odd, $\mathbf{C}$ is an even permutation and $\langle \mathbf{C}, (\mathbf{C}^{(k)})^d \rangle = A_N$. If N is even, $\mathbf{C}$ is an odd permutation and $\langle \mathbf{C}, (\mathbf{C}^{(k)})^d \rangle = S_N$. $\qquad \square$

Lemma 4.4. *In a deck of $N = nk + q$ cards, $0 \le q < k$, $\mathbf{O}_k$ is an odd or even permutation as $\left[\frac{1}{4}k(k - 1)n(n - 1) + \frac{1}{2}q(q - 1)(n + 2) + q(k - q)n\right]$ is an odd or even number.*

Proof. Recall the permutation matrix $P_{k,N}$ that corresponds to $\mathbf{O}_k$ on a deck of size $N = nk + q$:

$$P_{k,N} = \begin{bmatrix} E_{11} & E_{21} & \cdots & E_{q1} & F_{q+1,1} & \cdots & F_{k1} \\ \vdots & \vdots & & \vdots & \vdots & & \vdots \\ E_{1n} & E_{2n} & \cdots & E_{qn} & F_{q+1,n} & \cdots & F_{kn} \\ G_{1,n+1} & G_{2,n+1} & \cdots & G_{q,n+1} & 0 & \cdots & 0 \end{bmatrix},$$

where $E_{ij} \in \mathbb{R}_{k \times (n+1)}, F_{ij} \in \mathbb{R}_{k \times n}$, and $G_{ij} \in \mathbb{R}_{q \times (n+1)}$ are all unit matrices with 1 in the (i, j) position and zero elsewhere. The proof relies on a basic result from elementary group theory. We transform $P_{k,N}$ into the identity matrix $\mathbf{I}_N$ by shifting columns in the matrix while counting adjacent column transpositions. The parity of all these transpositions is equal to the parity of $\mathbf{O}_k$ as a permutation. First, we shift the last column of the first block-column to the right of the matrix to produce

$$\begin{bmatrix} F_{11} & E_{21} & \cdots & E_{q1} & F_{q+1,1} & \cdots & F_{k1} & 0 \\ \vdots & \vdots & & \vdots & \vdots & & \vdots & \\ F_{1n} & E_{2n} & \cdots & E_{qn} & F_{q+1,n} & \cdots & F_{kn} & 0 \\ & & & & & & & 1 \\ 0 & G_{2,n+1} & \cdots & G_{q,n+1} & 0 & \cdots & 0 & \vdots \\ & & & & & & & 0 \end{bmatrix}$$

The last column of the first block-column is shifted past the $q-1$ block-columns with $n+1$ columns and then past $k-q$ block-columns with n columns . Thus $(q-1)(n+1)+(k-q)n$ adjacent column transpositions are required to execute this shift.

Next, we shift the last column of the second block-column to the right of the matrix to yield

$$\begin{bmatrix} F_{11} & F_{21} & \cdots & E_{q1} & F_{q+1,1} & \cdots & F_{k1} & 0 & 0 \\ \vdots & \vdots & & \vdots & \vdots & & \vdots & \vdots & \vdots \\ F_{1n} & F_{2n} & \cdots & E_{qn} & F_{q+1,n} & \cdots & F_{kn} & 0 & 0 \\ & & & & & & & 1 & 0 \\ & & & & & & & 0 & 1 \\ 0 & 0 & \cdots & G_{q,n+1} & 0 & \cdots & 0 & \vdots & \vdots \\ & & & & & & & 0 & 0 \end{bmatrix}$$

The last column of the second block-column is shifted past the remaining $q-2$ block-columns with $n+1$ columns and then past $k-q$ block-columns with n columns and past the new rightmost column. Thus $(q-2)(n+1)+(k-q)n+1$ adjacent column transpositions are required for this shift.

Finally, the last column of the qth block-column is shifted to the right of the matrix for the total number of column transpositions equaling

$$(q-1)(n+1)+(k-q)n+0$$
$$+(q-2)(n+1)+(k-q)n+1$$
$$\cdots$$
$$+(q-q)(n+1)+(k-q)n+(q-1)$$
$$= \tfrac{1}{2}q(q-1)(n+1)+q(k-q)n+\tfrac{1}{2}q(q-1)$$
$$= \tfrac{1}{2}q(q-1)(n+2)+q(k-q)n.$$

The resulting matrix is column equivalent to $P_{k,N}$,

$$P_{k,N} = P_{k,nk+q} \widetilde{\text{col}} \begin{bmatrix} P_{k,nk} & 0 \\ 0 & \mathbf{I}_q \end{bmatrix},$$

where $\widetilde{\text{col}}$ indicates column equivalence.

The matrix $P_{k,nk}$ has a simpler block structure than $P_{k,nk+q}$:

$$P_{k,nk} = \begin{bmatrix} E_{11} & E_{21} & \dots & E_{k1} \\ E_{12} & E_{22} & \dots & E_{k2} \\ \vdots & \vdots & & \vdots \\ E_{1n} & E_{2n} & \dots & E_{kn} \end{bmatrix}.$$

Now we need only transform $P_{k,nk}$ into I_{nk} by column transpositions. We first take the left column of each block-column and shift it to the left of the matrix. No transpositions are required to shift the leftmost column of the first block-column, $n - 1$ adjacent column transpositions shift the first column of the second block-column, and finally $(k - 1)(n - 1)$ transpositions gets the first column of the kth block-column to the left. The resulting matrix is

$$P_{k,nk} \widetilde{\text{col}} \begin{bmatrix} \mathbf{I}_k & 0 & \dots & 0 \\ 0 & H_{12} & \dots & H_{k2} \\ \vdots & \vdots & & \vdots \\ 0 & H_{1,n-1} & \dots & H_{k,n-1} \end{bmatrix},$$

where $H_{ij} \in \mathbb{R}_{k \times (n-1)}$ are all unit matrices with 1 in the (i, j) position and zero elsewhere. A total of

$$1(n - 1) + 2(n - 1) + \cdots + (k - 1)(n - 1) = \tfrac{1}{2}k(k - 1)(n - 1)$$

adjacent column transpositions complete this part of the transformation.

We now shift the first column of each of the $n-1$ column-blocks containing the H_{ij} to the left of the block-column containing H_{1i}. No transpositions are required to shift the first column of the first block-column in place, $n-2$ adjacent column transpositions shift the first column of the second block-column, and finally $(k - 1)(n - 2)$ transpositions moves the first column of the kth block column to the left. A total of

$$1(n - 2) + 2(n - 2) + \cdots + (k - 1)(n - 2) = \tfrac{1}{2}k(k - 1)(n - 2)$$

adjacent column transpositions yield $\mathbf{I}_{2k}$ as the upper left block. Continuing in this manner, $P_{k,nk}$ is transformed into $\mathbf{I}_{nk}$ with

$$\tfrac{1}{2}k(k-1)(n-1)+\tfrac{1}{2}k(k-1)(n-2)+\cdots+\tfrac{1}{2}k(k-1)(n-2) = \tfrac{1}{4}k(k-1)n(n-1)$$

adjacent column transpositions.

Thus $P_{k,N}$ can be transformed into $\mathbf{I}_N$ with

$$\frac{1}{4}k(k-1)n(n-1) + \frac{1}{2}q(q-1)(n+2) + q(k-q)n$$

adjacent column transpositions. $\qquad\qquad\square$

Generalized Shuffle Groups

With the preliminary work behind us, we return to the original object of study: $\langle \mathbf{O}_k, \mathbf{C} \rangle$. Morris first described the group in his dissertation [**76**], and Morris and Hartwig came up with the proof given here [**74**].

Theorem 4.5. (Morris and Hartwig) *In a deck of $N = nk + q$ cards, $0 \le q < k$, the group $\langle \mathbf{O}_k, \mathbf{C} \rangle$ generated by the generalized out-shuffle $\mathbf{O}_k$ and the simple cut $\mathbf{C}$ is:*

(1) if $n = 0$ (and thus $N = q$), the cyclic group of order $N, \langle \mathbf{C} \rangle$;

(2) if $n \ge 1$ and $q = k - 1$, the semidirect product $\langle \mathbf{O}_k \rangle \times \langle \mathbf{C} \rangle$;

(3) if $n \ge 1$ and $q \ne k - 1$, and if N is odd and $\mathbf{O}_k$ is an even permutation, the alternating group A_N;

(4) if $n \ge 1$ and $q \ne k - 1$, and if N is even or $\mathbf{O}_k$ is an odd permutation, the symmetric group $\mathbf{S}_N$.

Proof.

(1) If $n = 0$, $N = 0k + q$, and $\mathbf{O}_k = $ id. Thus $\langle \mathbf{O}_k, \mathbf{C} \rangle = \langle \text{id}, \mathbf{C} \rangle = \langle \mathbf{C} \rangle$.

(2) If $n \ge 1$ and $q = k - 1$, then by Theorem 4.1, $\mathbf{O}_k\mathbf{C} = \mathbf{C}^k\mathbf{O}_k$. Following the same arguments as in Theorem 3.4, any product of the form $\mathbf{O}_k\mathbf{C}^a$ can be rewritten as $\mathbf{C}^{ka}\mathbf{O}_k$, and so $|\langle \mathbf{O}, \mathbf{C} \rangle| \le N \cdot o(\mathbf{O}_k, N)$. In particular, any element of $\langle \mathbf{O}_k, \mathbf{C} \rangle$ can be written as $\mathbf{C}^a\mathbf{O}_k^b$ for some a and b. The only

powers in common with $\mathbf{O}_k$ and $\mathbf{C}$ are

$$\mathbf{O}_k^{o(\mathbf{O}_k, N)} = \mathbf{C}^N = \text{id.}$$

Finally, if $|\langle \mathbf{O}_k, \mathbf{C} \rangle| < N \cdot o(\mathbf{O}_k, N)$, then there exist a, b, c, and d such that $\mathbf{C}^a \mathbf{O}_k^b = \mathbf{C}^c \mathbf{O}_k^d$, but there are only trivial solutions to $\mathbf{C}^{a-c} = \mathbf{O}_k^{d-b}$, so $|\langle \mathbf{O}_k, \mathbf{C} \rangle| = N \cdot o(\mathbf{O}_k, N)$. Again following Theorem 3.4, we show $\langle \mathbf{C} \rangle \lhd \langle \mathbf{O}_k, \mathbf{C} \rangle$ by showing $A\mathbf{C}^r A^{-1} \in \langle \mathbf{C} \rangle$ for any r and all $A \in \langle \mathbf{O}_k, \mathbf{C} \rangle$. Since any element of $\langle \mathbf{O}_k, \mathbf{C} \rangle$ can be written as $\mathbf{C}^a \mathbf{O}_k^b$, we have

$$A\mathbf{C}^r A^{-1} = \mathbf{C}^a \mathbf{O}_k^b \mathbf{C}^r \mathbf{O}_k^{-b} \mathbf{C}^{-a} = \mathbf{C}^a \mathbf{C}^{k^b r} \mathbf{O}_k^b \mathbf{O}_k^{-b} \mathbf{C}^{-a} = \mathbf{C}^{k^b r} = \langle \mathbf{C} \rangle,$$

so $\langle \mathbf{C} \rangle \lhd \langle \mathbf{O}_k, \mathbf{C} \rangle$. Finally $\langle \mathbf{C} \rangle \cap \langle \mathbf{O}_k \rangle = \text{id}, \langle \mathbf{O}_k \rangle \langle \mathbf{C} \rangle = \langle \mathbf{O}_k, \mathbf{C} \rangle$, and thus $\langle \mathbf{O}_k, \mathbf{C} \rangle$ is the semi-direct product of $\langle \mathbf{O}_k \rangle$ and $\langle \mathbf{C} \rangle$.

(3) If $n \geq 1$ and $q \neq k - 1$, then by Theorem 4.1, $\mathbf{O}_k \mathbf{C} = (\mathbf{C}^{(k)})^{q+1} \mathbf{O}_k$, so $(\mathbf{C}^{(k)})^{q+1} \in \langle \mathbf{O}_k, \mathbf{C} \rangle$. If $d = \gcd(k, q + 1)$, then not only is $(\mathbf{C}^{(k)})^d \in \langle \mathbf{O}_k, \mathbf{C} \rangle$, but also $\langle \mathbf{C}, (\mathbf{C}^{(k)})^d \rangle \subseteq \langle \mathbf{O}_k, \mathbf{C} \rangle$. Since $N \neq k$ and $q \neq k - 1$, by Lemma 4.3, $\langle \mathbf{C}, (\mathbf{C}^{(k)})^d \rangle$ is the alternating group A_N if N is odd, and hence $A_N \subseteq \langle \mathbf{O}_k, \mathbf{C} \rangle$. Thus $\langle \mathbf{O}_k, \mathbf{C} \rangle$ is A_N or S_N. Since N is odd, $\mathbf{C}$ is an even permutation, and if $\mathbf{O}_k$ is an even permutation, then both generators are even, so $\langle \mathbf{O}_k, \mathbf{C} \rangle = A_N$.

(4) If $n \geq 1$ and $q \neq n - 1$, then by the arguments in (3), $\langle \mathbf{O}_k, \mathbf{C} \rangle$ is A_N or S_N. If N is even, $\mathbf{C}$ is an odd permutation and hence $\langle \mathbf{O}_k, \mathbf{C} \rangle = S_N$, or if $\mathbf{O}_k$ is an odd permutation, $\langle \mathbf{O}_k, \mathbf{C} \rangle = S_N$.   □

Generalizing the In-Shuffle

So far our generalizations have been limited to the out-shuffle and have kept the top card on top. Now we must deal with the in-shuffle, and it is a much less pleasant creature to extend. In fact, this is an area with open questions for further research. The algebraic extension is the simplest. The ordinary out-shuffle, $(\mathbf{O}_2)$, and in-shuffle have the "nicest" definitions on an odd deck; both use the same modulus. Recall that on a deck of size $N = kn - 1$, $\mathbf{O}_k(p) \equiv kp \pmod{N - 1}$. This leads directly to the algebraic generalization of the in-shuffle.

Definition. The generalized perfect in-shuffle of order k, $\mathbf{O}_{k,r}, 0 \leq r < n$, is defined on a deck of $N = kn - 1$ cards as $\mathbf{O}_{k,r}(p) \equiv kp + r \bmod(kn - 1)$.

With this notation, $\mathbf{O}(p) = \mathbf{O}_{2,0}(p)$ and $\mathbf{I}(p) = \mathbf{O}_{2,1}(p)$. Once mathematicians start generalizing, everything is fair game. So let's consider the function $\delta : \{\mathbf{0}, \mathbf{I}\} \longrightarrow \{0, 1\}$ and its extension to generalized shuffles.

Definition. The function δ is a mapping from generalized perfect k-shuffles to the integers $0, \ldots, k - 1$.

$$\delta : \{\mathbf{O}_{k,0}, \mathbf{O}_{k,1}, \ldots, \mathbf{O}_{k,k-1}\} \longrightarrow \{0, 1, \ldots, k - 1\} \qquad \delta(\mathbf{O}_{k,r}) = r$$

Now we can nicely extend Theorem 2.6, the Fundamental Theorem of Faro-Shuffling on Odd Decks, which we state without proof. Medvedoff and Morrison give a proof for moving the top card to any position [68].

Theorem 4.6. (The Fundamental Theorem of Generalized Faro Shuffling)
In a deck of size $N = kn - 1$ cards, a sequence of t perfect k-shuffles, $\mathbf{S}_i, 1 \le i \le k$, moves the card in position p as follows:

$$\mathbf{S}_t \ldots \mathbf{S}_1(p) \equiv k^t + \sum_{i=1}^{t} k^{t-i} \delta(\mathbf{S}_i) \bmod(N).$$

Thus a k-handed magician could extend Alex Elmsley's method used in Trick 2.9 to move the top card to any position p in $\sim \log_k(p)$ generalized shuffles. Then again, if a magician had k hands, there are many more impressive things that could be done!

This algebraic extension produces k different k-shuffles, $\mathbf{O}_{k,0}, \ldots, \mathbf{O}_{k,k-1}$. When we look at physical generalizations, however, there are more in-shuffles, $k!$ in all. In the card game model of a generalized out-shuffle the deck is divided into k equal (or nearly-equal) piles. The first card of the first player becomes the top card of the deck, the first card of the second player becomes the next-to-the-top card, and so on, assuming the players play their cards in order from first to last. There are, however, $k!$ orders in which the players could agree to play their cards, and these correspond to the $k!$ in-shuffles. Things become very interesting when the deck is of size $N = nk + q$.

Morris and Hartwig generalized the in-shuffle for a deck of $N = nk + q$, not algebraically, but physically, from the card game model [74]. They defined $\mathbf{O}_{k,r}$ to be the permutation that results from cutting enough cards so the rth player has the top card on top of his packet, and then performing $\mathbf{O}_k$. Using the Morris and Hartwig definition, $\mathbf{O}_{k,r} = \mathbf{O}_{k,0}\mathbf{C}^x$ for some x, so $\langle \mathbf{O}_{k,r}, \mathbf{C} \rangle = \langle \mathbf{O}_{k,0}\mathbf{C}^x, \mathbf{C} \rangle = \langle \mathbf{O}_{k,0}, \mathbf{C} \rangle$.

Medvedoff and Morrison looked at the group generated by all $k!$ shuffles on a deck of size $N = kn$ [**68**]. Through computational methods they developed conjectures for the groups of 3-shuffles and 4-shuffles, but they found no general results.

Here are a few ideas for further research on generalized in-shuffles. Results on decks of size $N = nk + q$ may be more difficult than for decks of size nk or $nk - 1$.

- What is the most "natural" generalization of the in-shuffle? (This would be a definition that extends the most important theorems, though there could be more than one natural extension.)
- What is the order of a generalized in-shuffle?
- Are there closed form expressions for generalized in-shuffles?
- What is the group structure generated by various sets of shuffles, and what is the most "interesting" set of shuffles to study?
- Are there any good tricks using generalized shuffles?

Trick 4.7. ("The Triple Seekers") The magician performs the first two shuffles of a triple perfect shuffle, O_3, for an incomplete shuffle control with three packets (See Appendix 2). The top packet is riffled, spectator 1 says "Stop" and remembers the card. The top third is removed from the other two packets and placed on the table. The middle packet is riffled, spectator 2 says "Stop," and remembers the card. This third is removed from the remaining packet and placed on the table. The magician cuts the final packet in his hand and shows the face card, which will be the magician's card. This card is reversed, the deck reassembled, and any number of perfect shuffles and cuts are performed. Finally the magician performs a complete triple shuffle and spreads the cards. The cards to either side of the magician's reversed card are the cards of spectators 1 and 2.

Explanation. This is the only faro trick I've invented, and I like the mathematics of it better than any other faro trick. I only wish the intensity of the effect approached the appeal of its mathematics. A triple shuffle is just too tediously slow to make this a really dramatic trick for all audiences, but, for magicians or mathematicians (both of whom think they know something about what is going on), it has a unique impact.

The Selections and Control

1. *Perform an incomplete triple shuffle.* The size of the deck must be an odd multiple of 3. (I use a regular deck with one card left in the case—51 cards.) See Appendix 2 for details on performing a triple shuffle. Shuffle the top and bottom thirds of the deck into the middle third, pushing the packets together about $\frac{3}{4}$ inch. Hold the cards with their faces toward the left palm, with the left ring and little fingers free to hold a break in the bottom third.

2. *Riffle the upper corner of the top packet with the right index finger and have a spectator select a card.* The card chosen by the first spectator must be in the bottom half of the packet. You can force them to stop in the bottom half by riffling slowly and saying, "Just say Stop" when you want to select a card." When the command is issued, pull back the corner enough for the spectator to peek at the card (and also enough for a separation or "break" to occur in the bottom packet). Insert the tip of your left little finger in the break (figure 5). The top packet is pulled free from the other two woven-together packets and is placed on the table.

FIGURE 5.   As the right index finger pulls back the corner of the chosen card in the top third of the deck, the left little finger is inserted in the "break" in the bottom third.

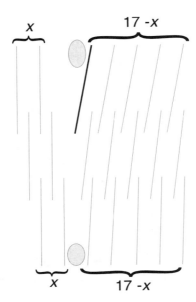

FIGURE 6.   The right index finger above and the left little finger below. The first chosen card is highlighted.

Recapitulation. The card of spectator 1 is in packet 1 on the table. We don't know where it's located, but we have preserved an exact mirror image of its location with the break held in the bottom packet. Assume there are x cards below the chosen card, so that it's the $17 - x$th card in the top packet, now sitting on the table. You're ready to have the next card selected.

3. *Riffle the upper corner of the top packet with the right index finger and have a spectator select a card.* The card chosen by the second spectator must be above the selection of the first spectator so you can put your left ring finger in the break. As with the first spectator, have the second say "Stop" to select a card. Pull back the corner at the command, let the spectator peek at the card, and create a break in the bottom packet. Insert the tip of your left ring finger in the break (figure 7). Pull the top packet free and place it on the table to the right of the first packet.
Recapitulation. The card of spectator 2 is in packet 2 on the table, to the right of the packet with the card of spectator 1. We don't know where the second card is located, but as with the first card, we have an exact mirror image of its location in the bottom packet. Assume there are y cards below

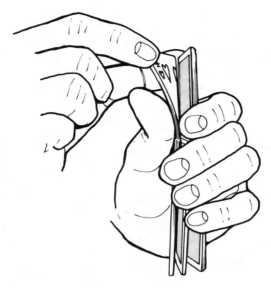

FIGURE 7. As the right index finger pulls back of the corner of the chosen card in the middle third of the deck, the left ring finger is inserted in the "break" in the bottom third of the deck.

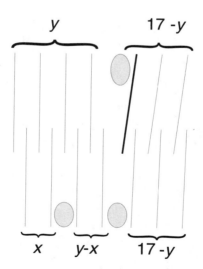

FIGURE 8. The right index finger above and the left little and ring fingers below. The second chosen card is highlighted.

the second chosen card, $x < y$, so that it's the $17 - y$th card in the middle packet. You are holding two breaks in the packet in your hand, at x and y from the bottom (figure 9). The little finger holds a break above x cards, and the ring finger holds a break at $y - x$ cards above the little finger break; there are $17 - y$ cards above the ring-finger break.

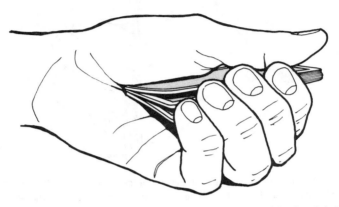

FIGURE 9.   The bottom third of the deck held in the left hand with a break held by the little and ring fingers.

4. *Choose your card and show it to the audience.* Cut off the $y - x$ cards above the ring-finger break and show the bottom card (figure 10). Make this cut in a casual way without letting the audience know you're holding a break. Announce, "I'll let this be my card," as your turn over the packet. Show the bottom card, $17 - y$ from the top of the packet, and put the packet face down to the right of the other two packets on the table. Casually cut at the little-finger break and put the $y - x$ cards on top of the middle packet. Put the remaining x cards on the left packet.

   Recapitulation. Here's the situation just after "choosing" your card and putting the packet on the table. Three packets are on the table: packet 1 on the left contains the card of spectator 1, $17 - x$ from the top; packet 2 in the center has the card of spectator 2, $17 - y$ from the top; packet 3 on the right packet has your card, the bottom one of $17 - y$. (In the diagrams that follow, the sub-packets with the chosen cards are shown in bold for clarity.) You are still holding y cards in your hand with a ring-finger break between $y - x$ and x cards.

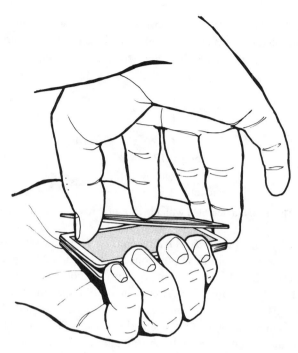

FIGURE 10.   Cutting off the $y - x$ cards above the ring-finger break. The bottom card of this packet will be the magician's "chosen" card.

17 − x	**17 − y**		top
x	y	**17 − y**	bottom
Packet 1	Packet 2	Packet 3	

Cut the cards remaining in your hand at the break, and put the $y - x$ cards on top of packet 2. Put the remaining x cards on packet 1; there are no cards left in your hands. All 51 cards are on the table in three packets.

x	$y - x$		top
17 − x	**17 − y**		
x	y	**17 − y**	bottom
Packet 1	Packet 2	Packet 3	

5. *Reverse your card and reassemble the deck.* Pick up the right packet and reverse the bottom card, your card. Put the packet in your left hand. Pick up the middle packet and put it on top of the packet in your hand. Pick up the last packet and put it on top of the other two.

Recapitulation. All the hard work is over—it's easy sailing from here. The reassembled deck has the three chosen cards exactly 17 cards apart in positions 16, 33, and 50. (Don't forget—the top card is in position 0).

$$
\begin{array}{ll}
\text{Packet 1} \left\{ \begin{array}{l} x \\ \mathbf{17-x} \\ x \end{array} \right\} & 17 \text{ cards} \\[2mm]
\text{Packet 2} \left\{ \begin{array}{l} y-x \\ \mathbf{17-y} \\ y \end{array} \right\} & 17 \text{ cards} \\[2mm]
\text{Packet 3} \left\{ \begin{array}{l} \mathbf{17-y} \end{array} \right\} & 17 \text{ cards}
\end{array}
$$

The deck reassembled with the chosen cards seventeen apart.

6. *Cut and faro shuffle the cards as many times as the spectator wants.* After a cut, the chosen cards will be at positions z, $17 + z$, and $34 + z$, for some z. After an out-shuffle, the cards are in positions $2z \pmod{51}$, $34 + 2z \pmod{51}$, and $68 + 2z \equiv 17 + 2z \pmod{51}$. After an in-shuffle, the cards are at positions $2z + 1 \pmod{51}$, $34 + 2z + 1 \pmod{51}$, and $17 + 2z + 1 \pmod{51}$. In other words, any combination of **O**'s, **I**'s, and **C**'s leave the three chosen cards exactly 17 apart. Don't overdo the shuffling or you'll bore the audience. As a practical matter, three shuffles with intervening cuts should be enough—if the audience isn't satisfied by this, then they're not going to be satisfied with another dozen shuffles and cuts.

The Discovery

7. *Perform a triple shuffle, O_3, and spread the cards.* A triple shuffle on a deck of $3n$ will bring together cards that are n apart. To insure that your reversed card is between those of spectators 1 and 2, use the final cut before the triple shuffle to bring the reversed card to the middle third of the deck. The reversed card is easy to follow because it stands out in the deck, and it's easy to get it to the middle third of the deck with a cut. After the triple shuffle, all that's left is showmanship.

8. *Remove the reversed card and the two on either side, turn them over, and show them to the startled spectators as their eyes bulge and jaws drop.* Well, this is the way I like to remember it.

5
Dynamic Computer Memories

I was with a few friends when one of them brought out a deck of cards. He said, "Would you like to see a card trick I learned?" I said, "Sure," and he had me choose a card.

I looked at the card, returned it to the deck, and he said he would lose the card with a special "unshuffle." He cut the cards and shuffled them—it looked like they went together exactly every-other-one.

My friend said the deck was being "unshuffled," and he told me to look at the edge. I could just read the letters U-N-S-H-U-F-F-L-E-D on the side of the deck, repeated four times. He did another shuffle and the letters appeared larger, repeated only twice this time. After a final shuffle, there was one set of letters, large and clear.

He spread the cards and showed me that the deck was back in its original order from the factory, truly unshuffled. He then asked me what my card was. I said, "The eight of hearts." He showed me the other side of the deck, and clearly spelled out was "Eight of Hearts."

How did he do that?

Memory is a vitally important part of any computer system. A recent advertisement for a computer system hawks "133 MHZ Pentium® Processor, 16

MB RAM/1.08 GB Hard Drive"; right after the processor specifications come those of the memory.[1] Most computer systems offer two types of memory, fast, expensive random access memory on a chip, and slow, cheap bulk memory on a hard disk. This seems to be a nice marriage of different needs in a computer system: blazing speed for immediate access to data, and almost unlimited capacity for long-term storage.

Harold Stone coined the term *dynamic memory* to describe memories where "the storage technology inherently requires that there be a continuous circulation of data." [**93**] For example, a hard or floppy disk rotates to access its data, and the data on each track is accessible only at one location—under the read head. In contrast we could call a random access memory a *static memory*, because each datum remains fixed in a single location, but every storage location can be accessed instantly because each is hard-wired to the "outside."

The initial motivation for studying dynamic memories came from a computer technology change in the early 1970s. Manufacturers began to abandon magnetic core memories for smaller, faster semiconductor ones; magnet cores retained their states when power was turned off, but semiconductors "forgot" everything. One alternative to volatile, "forgetting" semiconductor memory was the magnetic bubble memory, which routed "bubbles" in a magnetic field on a crystal plane by causing them to follow paths laid out on the crystal surface. The goal of much of that then-fashionable research was to find suitable paths and control algorithms. As things developed, magnetic bubbles vanished (popped?), and fast, cheap disks and low-power memory with battery back-up now provide the nonvolatile storage.

Stone led much of the research on bulk memories; he thought these would probably be integrated circuits, possibly using magnetic bubbles, charge-coupled diodes, or transistors. He speculated such devices are likely to have the following characteristics [**94**].

1. *Low internal connectivity*: Each internal data storage cell will connect to few other cells.

2. *Limited number of read/write ports*: Each memory will have only one or two read/write ports, and data must be routed physically from an internal storage location to a read/write port to be accessed.

---

[1] Perhaps no other item will date this book more than these system specifications, taken from the front of a computer supplies catalog. By the time this manuscript wends its way to the reader's hands, the Pentium processor may be virtually obsolete and 16 MB RAM so woefully inadequate as to cause snickers. For the curious, this note was written January 24, 1997.

3. *Dynamic storage*: A datum with a specific logical address need not be stored in any particular physical storage location, and it may change its physical storage location many times during its storage lifetime. To access a datum with a given logical address, the memory determines the physical location of that datum at that time, and routes the datum to a read/write port, as shown in figure 1.

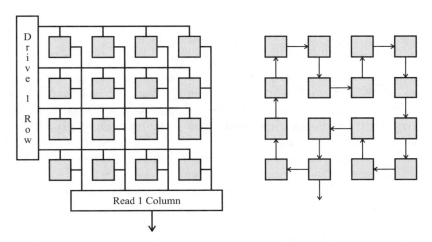

FIGURE 1.   A static memory on the left, with each cell having a path off the chip, and a dynamic memory on the right, with a single read/write port connected off the chip. Note how much more space is occupied by interconnections in a static memory.

The Shift-Register Memory

To better understand these concepts, consider a cyclic shift register as a single track on a disk. Each cell in the register stores a single datum, perhaps in a transistor flip-flop or a charge-coupled diode, and all cells but one are connected to only the next cell in a large circle. The exception is a cell connected additionally to the read/write port. When the register is "clocked," that is, when each cell receives a signal to discharge its contents, the data shifts around the register. A datum is accessed by shifting it to the read/write port (figure 2).

The problem with such a design is the control: deciding how many times to shift the memory to access a datum. The shift-register memory in figure 2 implements the inverse simple cut $\mathbf{C}^{-1}$, moving the datum at location p to location $p + 1$, or cutting a single datum ("card") from the bottom to the top

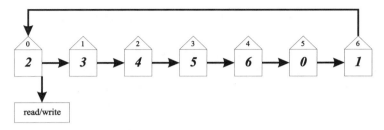

FIGURE 2. A shift-register memory with the physical storage location shown on top of each cell in the register, and the logical address shown inside.

of the shift register memory ("deck"). Note that in figure 2 logical address 2 is in physical location 0, the read/write port. In figure 2 each logical address l is in physical location $p \equiv l - 2 (\mathrm{mod}\, 7)$. In general, if r/w is the current content of the read/write port then the physical location of each logical address is $p \equiv l - r/w (\mathrm{mod}\, N)$, where N is the size of the memory. In a shift-register memory with interconnection $\mathbf{C}^{-1}$, if a datum is in physical location p, a shift of $-p (\mathrm{mod}\, N)$ moves it to the read/write port.

The value of r/w can be tracked with a mod N counter that is decremented each time the memory is clocked (shifted). The computation $l - r/w$ is made in a small arithmetic logic unit (ALU). If the memory is of size N, then the ALU and counter are of size $o(\log_2 N)$, because it only takes $\log_2 N$ bits to represent a number of size N. The accessing algorithm for a shift-register memory is rather simple.

Data Accessing Algorithm for a Shift-Register Memory

1. INPUT: l = logical address to access to read/write port.
2. COMPUTE: the physical address $p \equiv l - r/w (\mathrm{mod}\, N)$.
3. SHIFT: $-p (\mathrm{mod}\, N) \equiv$ the amount to shift the register.
4. DECREMENT: $r/w \equiv r/w - p (\mathrm{mod}\, N)$.

This is illustrated in figure 3.

The cyclic interconnection of the shift-register memory is the simplest possible form of a dynamic memory. The average time to access any randomly chosen datum is called the mean random access (MRA). It's easy to see that for a shift-register memory the best case time is 0, the worst case is $N - 1$,

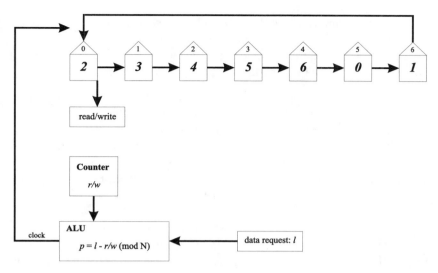

FIGURE 3. Shift-register memory with control mechanism: arithmetic logic unit and counter. The data request, l, and the current contents of the read/write port, r/w, are input to the ALU, and their difference, p, is computed mod N. The memory is shifted $-p(\bmod N)$ times, and the counter decremented p times.

and MRA $= \frac{1}{2}(N - 1)$. We can imagine a dynamic memory with r data paths connecting the cells. At each clock time one of the r permutations is selected and applied to the data in the memory. With a good control algorithm, such a robustly connected memory should have faster random access. (It certainly wouldn't be any slower.) Stone addressed this issue when he defined the concept of a dynamic memory.

Lemma 5.1. (Stone) *A lower bound on the minimum worst case access time for a dynamic memory with r interconnection patterns is $M(r,N)$, where $M(r,N)$ is the smallest integer that satisfies*

$$\frac{r^{M(r,N)+1} - 1}{r - 1} \geq N.$$

Proof. At each clock time we choose one of r paths for the data to follow. Thus in one clock time we can access one of r data, in two clock times r^2 data, and so on. In M clock times we can access no more than

$$\sum_{i=0}^{M} r^i = \frac{r^{M+1} - 1}{r - 1}.$$

When this sum is less than N, there must be some datum that is inaccessible. When this sum is greater than or equal to N, there is the possibility we can access any datum, though we don't know the sequence of permutations to do this. Thus we must be at least that big. This proves the lemma.   □

As the size of the memory N gets large, the lower bound for the worst-case access time is approximately $\log_r N + \log_r (r - 1)$. For practical-sized memories two interconnections paths are more than enough. Consider a dynamic memory with $2^{20} = 1{,}048{,}576$ data items. (If each datum is a byte, then this is megabyte memory.) With one data path, a cyclic interconnection, the MRA $= 524{,}287.5$. With two data paths the best worst-case access time is 20. With three data paths the best worst-case is 13, faster than two paths, but nothing compared with the speed-up from one to two paths.

The cyclic interconnection of a shift-register memory has MRA $= \frac{1}{2}(N - 1)$. Stone's Lemma tells us that the right pair of interconnections could give random access in no worse than $\log_2 N$. Can we find two paths that shift data in our memory and achieve Stone's lower bound? Yes, we can! (And since this book is devoted to the perfect shuffle, it a good guess that the paths will be the out- and in-shuffles!) Stone's solution, slightly modified, is the perfect-shuffle memory.

The Perfect-Shuffle Memory

The out- and in-shuffles can be used as interconnections to dramatically improve the performance of a shift-register memory. Consider a memory with $N = 2^n - 1$ storage locations interconnected with two paths, an out-shuffle and an in-shuffle. As with the shift-register memory, we keep track of r/w, the logical address of the current content of the read/write port. Assuming the data is only permuted by shifts as with a shift-register memory, each logical record l is in physical location $p = l - r/w$, and a shift of $-p(\bmod N)$ will access it to the read-write port. The shift will be between 0 and $2^n - 1$ and can be expressed as a binary number with $n = \lceil \log_2 N \rceil$ bits. Since $2^n \equiv 1(\bmod N)$, Corollary 2.7 assures us that any n shuffles is equivalent to some shift of the memory. In particular, the n shuffles corresponding to the n binary digits of $-p(\bmod N)$, from most significant to least significant, with a 1 corresponding to an in-shuffle and a 0 to an out-shuffle, will shift the memory by $-p(\bmod N)$.

Consider the perfect shuffle memory in figure 4, and the process for accessing data item 1. The data request is sent to the ALU, which calls for the

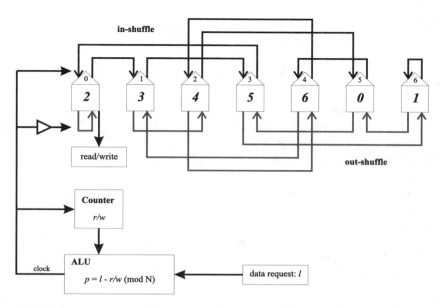

FIGURE 4. A perfect-shuffle memory with $r/w = 2$. To access logical record l at position $p \equiv l - r/w \pmod{N}$, the memory must be shifted to the right by $-p \pmod{N}$. To access datum 1 in position 6, shift the memory to the right by $1 \equiv -6 \pmod{7}$.

current value of r/w, and then computes $1 = -(1 - 2) \pmod{7}$, the right shift required to access data item 1 to location 0, the read/write port. The shift 1 is stored in a register as the binary number 001. These digits are then shifted out, one at a time, most significant digit first. If the digit is a 1, the in-shuffle path is clocked, and the data moves accordingly; if the digit is a 0, the out-shuffle path is clocked when the signal is inverted. The counter here is more sophisticated than a simple one-up mod N counter of a shift-register memory. The perfect shuffle memory counter recognizes the clock pulses as a binary number, and after n pulses increments itself accordingly.

Let's follow what happens as the memory in figure 4 accesses datum 1.

	Counter	Memory
0. The initial state of the memory.	$2_{10} = 010_2$	$[2, 3, 4, 5, 6, 0, 1]$
1. Data item 1 is requested, and the ALU computes $1 \equiv -(1 - 2) \pmod{7}$. The shift	$2_{10} = 010_2$	$[2, 3, 4, 5, 6, 0, 1]$

	Counter	**Memory**
will be 1, which is stored as $001_2 = 1_{10}$.		
2. The most significant bit is shifted out, a 0, which clocks the out-shuffle path.	$2_{10} = 010_2$	$[2, 6, 3, 0, 4, 1, 5]$
3. The next bit is shifted out, another 0, and another out-shuffle follows.	$2_{10} = 010_2$	$[2, 4, 6, 1, 3, 5, 0]$
4. The last bit is a 1, which results in an in-shuffle, and datum 1 is at the read/write port. Note that the memory is in a cyclic shift of its original state.	$2_{10} = 010_2$	$[1, 2, 3, 4, 5, 6, 0]$
5. Now that $3 = \log_2 N$ clock pulses have been sent, the counter decrements itself, $1 = 2 - 1$.	$1_{10} = 001_2$	$[1, 2, 3, 4, 5, 6, 0]$

By having two permutations in the memory, the out- and in-shuffles, we can achieve Stone's lower bound for randomly accessing data in the fastest possible time, $\log_2 N$. The time savings for a meg of memory is from an MRA of ~500,000 to a guaranteed access time of 20. This design is not without problems, however. Computer data is usually not accessed one datum at a time, but rather in large sequential blocks. The first datum of the block is accessed, and then each element in sequence. A perfect-shuffle memory elegantly solves the random access problem with $\log_2 N$ shuffles, but sequential access is accomplished by individually, albeit elegantly, accessing data at a time cost of $\log_2 N$ each.

For random access, a $2^{20} - 1$ perfect-shuffle memory beats a similarly-sized shift-register memory about 20 to 500,000. Sequential access, however, is a much different matter. In a shift-register memory, once the first element of a block is accessed, each subsequent block element takes only one delay. Thus a block of 50,000 takes about 550,000 shifts—about 500,000 (on average) to access the first element and 49,999 to access the rest of the block. The cyclic shift is perfectly designed to access sequential elements in a block. A perfect-shuffle memory takes exactly 20 permutations to access each element of the block, for a total delay of 1,000,000.

The Shift-Shuffle Memory

Alfred Aho and Jeffrey Ullman saw the sequential-access problem with a perfect-shuffle memory, and proposed a memory design that takes a little longer to randomly access the first datum in a block, but gives sequential access in unit time [**2**]. Their design, slightly modified, is the shift-shuffle memory shown in figure 5. Its accessing algorithm is based on a result in Theorem 3.5, $\mathbf{O} = \mathbf{CI}$ in odd decks. The shift-shuffle memory is of size $N = 2^n - 1$ and has two interconnections: the in-shuffle, $\mathbf{I}$, and simple cut, $\mathbf{C}$. The perfect-shuffle memory accessing algorithm is used to randomly access the first datum in a block, but $\mathbf{CI}$ replaces $\mathbf{O}$. Then $\mathbf{C}$ is used to sequentially access the remaining data in the block.

Let's follow the memory in figure 5 as it accesses a block of data beginning with item 1.

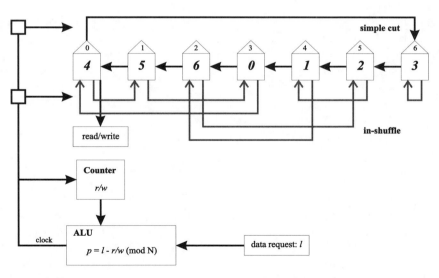

FIGURE 5. A shift-shuffle memory. The random access of the perfect-shuffle memory is used, except that when an out-shuffle permutation is needed, $\mathbf{CI}$ is substituted. Sequential access is achieved with $\mathbf{C}$.

	Counter	Memory
0. The initial state of the memory.	$4_{10} = 100_2$	$[4, 5, 6, 0, 1, 2, 3]$
1. Data item 1, the first in the block, is requested, and the ALU computes $3 \equiv -(1 - 4)(\mod 7)$. The shift will be 3, which is stored in the ALU as $011_2 = 3_{10}$.	$4_{10} = 100_2$	$[4, 5, 6, 0, 1, 2, 3]$
2. The most significant bit is shifted out, a 0, calling for an out-shuffle. (Remember O = CI.) First, an in-shuffle permutes the data.	$4_{10} = 100_2$	$[0, 4, 1, 5, 2, 6, 3]$
3. To complete the out-shuffle, a simple cut follows.	$4_{10} = 100_2$	$[4, 1, 5, 2, 6, 3, 0]$
4. The next bit is shifted out, a 1, and an in-shuffle follows.	$4_{10} = 100_2$	$[2, 4, 6, 1, 3, 5, 0]$
5. The last bit is a 1, which results in an in-shuffle, and datum 1 is at the read/write port. Note that the memory is in a cyclic shift of its original state.	$4_{10} = 100_2$	$[1, 2, 3, 4, 5, 6, 0]$
6. Now that $3 = \log_2 N$ clock pulses have been sent, the counter decrements itself, $1 \equiv 4 - 3(\mod 7)$.	$1_{10} = 001_2$	$[1, 2, 3, 4, 5, 6, 0]$
7. A simple cut accesses the next datum in the block. Because of greater complexity in the control algorithm for a shift-shuffle memory, the counter now increments itself, $2 \equiv 1 + 1(\mod 7)$.	$2_{10} = 010_2$	$[2, 3, 4, 5, 6, 0, 2]$

If we assume that an in-shuffle takes as long as a simple cut to perform on a memory chip, then the longest delay for a random access in a shift-shuffle memory is $2^n - 1$, which occurs when the memory is shifted by $1_{10} = 00 \ldots 01_2$. Two shifts will never occur—0 and $2^n - 1$—since each leaves the memory unchanged. If we assume that all shifts, $1, \ldots, 2^n - 2$, for random accessing are equally likely, then a shift-shuffle memory of size $N = 2^n - 1$ has MRA $= 1.5n$.

Details, Details, Details

The control algorithm for the perfect-shuffle memory (as well as the shift-shuffle memory) is essentially Elmsley's method for moving the top card to any location in the deck (Corollary 2.8 and Trick 2.9). Aho, Stone, and Ullman developed their algorithms with no idea that a perfect shuffle could be achieved with a regular deck in real time. Elmsley, on the other hand, seems to have thought the most practical application of his invention was to entertain an audience. This is a fascinating example of independent, parallel development of the same idea, and none of these clever inventors knew of the eighteenth-century experience with the perfect shuffle.

The perfect-shuffle and shift-shuffle memory random accessing algorithm is particularly elegant; calculating the location of the desired datum gives the control sequence. Minimal computation and storage is required to control the memory. As elegant as the control algorithm may be, the memories are not without a serious drawback—they are odd-sized; their storage capacity cannot be a power of two. Theoreticians love to solve the most general case; industry wants what will sell. (You don't see many ads for computers offering "8 MB -1 RAM.") Aho and Ullman sought to avoid this problem by putting a single dedicated storage cell on the chip for the the 2^nth data item with direct access to the read/write port; when it was to be accessed, the control would send it directly off the chip. For a large memory chip—perhaps a meg, 2^{20}—a single storage cell with direct access off the chip wouldn't add much to the control costs.

The ideal solution would be to find (1) a pair of permutations for a memory of size $N = 2^n$ (2) with random access in time $n = \log_2 N$ and (3) with sequential access in unit time. With a modification of its accessing algorithm, the perfect-shuffle memory achieves (1) and (2), and has random access in an average time of $1.5n$.

The Perfect-Shuffle Memory for $N = 2^n$

It is interesting to follow the development of dynamic memories in the pages of the *IEEE Transactions on Computers*. Stone first proposed the perfect-shuffle memory for memories of size $N = 2^n$ in 1972, but he did not address sequential accessing [93]. Aho and Ullman modified Stone's design in 1974 to yield the shift-shuffle memory for $N = 2^n - 1$, which solved the sequential accessing problem at the cost of using only odd-sized memories [2]. Stone realized how to get sequential accessing in memories of size 2^n in 1975 [**94**]. S. Brent Morris, Arthur Valliere, III, and Richard A. Wisniewski refined and generalized Stone's design in 1979 [**78**], [**79**].

Stone's algorithm for random accessing data with a perfect-shuffle memory of size 2^n is considerably different from that for odd-sized memories. Recall that the congruences defining out- and in-shuffles use different moduli in even memories; there is no known simple way to shift every element by a constant amount, as in Corollary 2.7. Before we explain an algorithm for random access, we need some additional notation.

Definition. For any number x, $b_n(x) = (x_{n-1}, x_{n-2}, \ldots, x_2, x_1, x_0)$ is its representation as an n-digit (or n-bit) binary number. For example, $b_5(13) = (0, 1, 1, 0, 1)$.

Definition. For any number x, $\rho(x)$ is the number equal to an end-around right shift of its binary representation, that is, $b_n(\rho(x)) = (x_0, x_{n-1}, x_{n-2}, \ldots, x_2, x_1)$. For example, $\rho(13) = 22$; $b_5(\rho(13)) = (1, 0, 1, 1, 0)$.

Definition. For any numbers x and y, $x \oplus y$ is the component-wise or bit-by-bit mod 2 sum of x and y as n-digit binary numbers, that is, $b_n(x \oplus y) = (x_{n-1} \oplus y_{n-1}, \ldots, x_1 \oplus y_1, x_0 \oplus y_0)$. For example, $13 \oplus 7 = 10$; $b_5(13 \oplus 7) = (0 \oplus 0, 1 \oplus 0, 1 \oplus 1, 0 \oplus 1, 1 \oplus 1) = (0, 1, 0, 1, 0)$.

The operations ρ and $\oplus$ are easy to implement in computer hardware and are often fundamental operations of a CPU. Note that $\rho(x \oplus y) = \rho(x) \oplus \rho(y)$. We now modify the definition of out- and in-shuffles in decks of size 2^n using the operations ρ and $\oplus$.

Lemma 5.2. *The out-shuffle on a deck of $N = 2^n$ cards, $0, \ldots, 2^n - 1$, is the permutation that moves the card in position p to position $\mathbf{O}(p)$ where $b_n(\mathbf{O}(p)) = \rho^{-1}(p)$.*

Proof. Case 1: $0 \le p \le 2^{n-1} - 1$. The binary representation of each number in this range has a 0 for its most significant digit, $p = (0, p_{n-2}, \ldots, p_1, p_0)$. From the definition of an out-shuffle, $\mathbf{O}(p) \equiv 2p \pmod{N - 1}$, and for p in the range $0 \le p \le 2^{n-1} - 1$ we do not need to consider congruences since $2p < 2^n - 1$, so $\mathbf{O}(p) = 2p$. Thus $b_n(\mathbf{O}(p)) = b_n(2p) = (p_{n-2}, \ldots, p_1, p_0, 0) = \rho^{-1}(p)$.

Case 2: $2^{n-1} \le p \le 2^n - 1$. The binary representation of each number in this range has a 1 for its most significant digit, $p = (1, p_{n-2}, \ldots, p_1, p_0)$, so that $2p = 2[(1,0,\ldots,0) + (0,p_{n-2},\ldots,p_0)]$. But $2(1,0,\ldots,0) = 2^n \equiv 1 \pmod{N-1}$ and $2(0,p_{n-2},\ldots,p_0) = (p_{n-2},\ldots,p_0,0)$. Thus $b_n(\mathbf{O}(p)) = b_n(1 + (p_{n-2},\ldots,p_0,0)) = \rho^{-1}(p)$. $\qquad\square$

Lemma 5.3. *The in-shuffle on a deck of $N = 2^n$ cards, $0,\ldots,2^n - 1$, is the permutation that moves the card in position p to position $\mathbf{I}(p)$, where $b_n(\mathbf{I}(p)) = \rho^{-1}(p) \oplus 1$.*

Proof. Similar to that of Lemma 5.2. $\qquad\square$

Corollary 5.4. *A perfect shuffle $\mathbf{S}$, either in or out, on a deck of $N = 2^n$ cards, $0,\ldots,2^n - 1$, is the permutation that moves the card in position p to position $\mathbf{S}(p)$, where $b_n(\mathbf{S}(p)) = \rho^{-1}(p) \oplus \delta(\mathbf{S})$.*

We now take a slightly different view of how a shuffle affects a deck. Rather than consider where the card in position p is moved, we derive a formula for what is moved into position p. We also adopt new notation to reflect this new view. Each position can be thought of as a "window" that gives us a peek into the arrangement of the deck. Rather than talk about the card in position k, we shall refer to the contents of window k, w_k. If we are interested in w_k at some specific time t, we use the notation $w_{k,t}$. We now set the stage for the random accessing algorithm for perfect-shuffle memories of size 2^n.

Lemma 5.5. *Assume a deck of $N = 2^n$ cards is in initial order at time $t = 0$, that is, $w_{k,0} = k$, and has been perfectly shuffled t times. Then*

$$w_{i,t} = w_{0,t} \oplus \rho^t(i).$$

Proof. By induction on t.

(1) We first show that $w_{i,1} = w_{0,1} \oplus \rho(i)$. There are two cases: $\mathbf{S} =$ Out and $\mathbf{S} =$ In.

   $\mathbf{S} =$ Out: Because the deck is out-shuffled, $w_{0,1} = w_{0,0} = 0$. By Lemma 5.2, the out-shuffle moves the card in position i to position $\mathbf{O}(i) = \rho^{-1}(i)$, so the card that moves to i at time 1 must have been in position $\rho(i)$ at time 0, as $\rho^{-1}\rho(i) = i$. Thus $w_{i,1} = \rho(i)$, and since $w_{0,1} = 0$, $w_{i,1} = w_{0,1} \oplus \rho(i)$.

S = In: Because the deck is in-shuffled, $w_{0,1} = 2^{n-1} = \rho(1)$. By Lemma 5.3, $\mathbf{I}(i) = \rho^{-1}(i) \oplus 1$, so the card that moves to position i at time 1 must have been in position $\rho(i \oplus 1)$ as $\rho^{-1}\rho(i \oplus 1) \oplus 1 = i \oplus 1 \oplus 1 = i$. Thus $w_{i,1} = \rho(i \oplus 1) = \rho(i) \oplus (1) = w_{0,1} \oplus \rho(i)$.

(2) For the induction hypothesis, we assume $w_{i,t} = w_{0,t} \oplus \rho^t(i)$.

(3) We now prove $w_{i,t+1} = w_{0,t+1} \oplus \rho^{t+1}(i)$. As in step (1), there are two cases: **S** = Out and **S** = In.

S = Out. The card that moves to position i at time $t + 1$, $w_{i,t+1}$, was in position $\rho(i)$ at time t, but this card is

$$w_{i,t+1} = w_{\rho(i),t} = w_{0,t} \oplus \rho^t(\rho(i)) = w_{0,t} \oplus \rho^{t+1}(i).$$

Since an out-shuffle was performed, $w_{0,t} = w_{0,t+1}$. Thus

$$w_{i,t+1} = w_{0,t+1} \oplus \rho^{t+1}(i).$$

S = In. The card that moves to position i at time $t + 1$ was in position $\rho(i \oplus 1)$ at time t, or in other words

$$w_{i,t+1} = w_{\rho(i\oplus 1),t} = w_{0,t} \oplus \rho^t(\rho(i \oplus 1)) = w_{0,t} \oplus \rho^{t+1}(i \oplus 1)$$
$$= w_{0,t} \oplus \rho^{t+1}(i) \oplus \rho^{t+1}(1).$$

Since an in-shuffle was performed,

$$w_{0,t+1} = w_{2^{n-1},t} = w_{\rho(1),t} = w_{0,t} \oplus \rho^t(\rho(1)) = w_{0,t} \oplus \rho^{t+1}(1).$$

Thus

$$w_{i,t+1} = w_{0,t+1} \oplus \rho^{t+1}(i). \qquad \square$$

This theorem is powerful: The current card in any position i is completely determined by the card in position 0, the number, t, of shuffles performed, and t end-around right shifts of i. Table 1 illustrates the application of Lemma 5.5. In practice it's only necessary to keep track of $t \pmod{n}$, since $\rho^x = \rho^{x+n}$.

Lemma 5.5 is a special case of a more general result from [**78**] that says the current card in position i is completely determined by the card in position k, the number, t, of shuffles performed and t end around right shifts of $i \oplus k$.

Lemma 5.6. (Morris, Valliere, and Wisniewski) *Assume a deck of $N = 2^n$ cards is in initial order at time $t = 0$ and has been perfectly shuffled t times.*

TABLE 1 A deck of $2^3 = 8$ cards perfectly shuffled from initial order, $w_{i,t} = w_{0,t} \oplus \rho^t(i)$

Initial			In			Out			Out		
i	$w_{i,0}$	$\rho^0(i)$	i	$w_{i,1}$	$\rho^1(i)$	i	$w_{i,1}$	$\rho^2(i)$	i	$w_{i,1}$	$\rho^3(i)$
0	0	000	0	4	000	0	4	000	0	4	000
1	1	001	1	0	100	1	6	010	1	5	001
2	2	010	2	5	001	2	0	100	2	6	010
3	3	011	3	1	101	3	2	110	3	7	011
4	4	100	4	6	010	4	5	001	4	0	100
5	5	101	5	2	110	5	7	011	5	1	101
6	6	110	6	7	011	6	1	101	6	2	110
7	7	111	7	3	111	7	3	111	7	3	111

Then

$$w_{i,t} = w_{k,t} \oplus \rho^t(i \oplus k).$$

To finish the foundation for our random-accessing algorithm for perfect-shuffle memories of size 2^n, we need a way to quickly determine $w_{0,t}$, from which we can in turn derive $w_{i,t}$.

Theorem 5.7. *(M-V-W) Assume of deck of $N = 2^n$ cards starts in initial order and has been perfectly shuffled t times. Then after another perfect shuffle **S**,*

$$w_{0,t+1} = w_{0,t} \oplus \delta(\mathbf{S})\rho^{t+1}(1).$$

Proof. By induction on t. Recall that $\delta(\mathbf{O}) = 0$ and $\delta(\mathbf{I}) = 1$.

(1) $\mathbf{S} = $ Out. $w_{0,1} = 0 = w_{0,0} = w_{0,0} \oplus \delta(\mathbf{O})\rho(1)$.

$\mathbf{S} = $ In. $w_{0,1} = 2^{n-1} = \rho(1) = w_{0,0} \oplus \delta(\mathbf{I})\rho(1)$.

(2) Assume $w_{0,t} = w_{0,t-1} \oplus \delta(\mathbf{S})\rho(1)$.

(3) $\mathbf{S} = $ Out. $w_{0,t+1} = w_{0,t} = w_{0,t} \oplus \delta(\mathbf{O})\rho^{t+1}(1)$.

$\mathbf{S} = $ In. $w_{0,t+1} = w_{2^{n-1},t} = w_{\rho(1),t}$.

By Lemma 5.6, $w_{\rho(1),t} = w_{0,t} \oplus \rho^t(\rho(1)) = w_{0,t} \oplus \rho^{t+1}(1) = w_{0,t} \oplus \delta(\mathbf{I})\rho^{t+1}(1)$.

$\square$

The work of Morris, Valliere, and Wisniewski [78] generalized all of Stone's results as far as possible. Theorem 5.7, like Lemma 5.5, is a special case of a more general result.

Theorem 5.8. (M-V-W) *Assume a deck of* $N = 2^n$ *cards starts in initial order and has been perfectly shuffled t times. After another perfect shuffle* **S**,

$$w_{i,t+1} = w_{i,t} \oplus \rho^t(i) \oplus \rho^{t+1}(i) \oplus \delta(\mathbf{S})\rho^{t+1}(1).$$

Corollary 5.9. (M-V-W) *Assume a deck of* $N = 2^n$ *cards starts in initial order and has been perfectly shuffled t times. After another k perfect shuffles,*

$$w_{0,t+k} = w_{0,t} \oplus \bigoplus_{i=1}^{k} \delta(\mathbf{S}_i)\rho^{t+1}(1).$$

We now have the results in place to prove the theorem that is the basis for random accessing in perfect shuffle memories of size 2^n.

Theorem 5.10. (M-V-W) *Assume a deck of* $N = 2^n$ *cards starts in initial order and has been perfectly shuffled t times. Then for any card d there exists an n-long sequence,* $\mathbf{S}_1, \ldots, \mathbf{S}_n$ *of perfect shuffles such that, when applied to the deck,*

$$w_{0,t+n} = d.$$

Proof. By Corollary 5.9, after n perfect shuffles, $\mathbf{S}_1, \ldots, \mathbf{S}_n$

$$w_{0,t+n} = w_{0,t} \oplus \bigoplus_{i=1}^{n} \delta(\mathbf{S}_i)\rho^{t+i}(1),$$

and so

$$w_{0,t+n} \oplus w_{0,t} = \bigoplus_{i=1}^{m} \delta(\mathbf{S}_1)\rho^{t+i}(1).$$

The value of $w_{0,t}$ is known, and we want $w_{0,t+n} = d$. Since $t + i(\mathrm{mod}\, n)$ runs through all values $0, \ldots, n - 1, \{\rho^{t+i}(1)|0 \le i \le n - 1\}$ contains each n-digit binary number with a single 1, and some linear sum of these numbers equals $w_{0,t} \oplus w_{0,t+n}$. We look at $\rho^{t+i}(1) \wedge w_{0,t} \wedge w_{0,t+n}$, the bit-by-bit application of the logical AND operator. If the result is 0, that is, $w_{0,t}$ and $w_{0,t+n}$ differ in the bit where $\rho^{t+i}(1)$ has a 1, then we let $\mathbf{S}_i = \text{In}$; otherwise $\mathbf{S}_i = \text{Out}$. This sequence of shuffles gives $w_{0,t} = d$. □

Random Accessing Algorithm for a Perfect-Shuffle Memory

1. INPUT: l = logical address to access to read/write port 0.
2. FOR $i = 1$ TO n:
   IF $\rho^{t+i}(1) \wedge w_{0,t} \wedge l = 0$ THEN IN-SHUFFLE
   ELSE OUT-SHUFFLE

TABLE 2   For a Perfect-Shuffle Memory of size $N = 2^3 = 8$, random accessing $l = 6$ to read/write port 0, after the shuffles **I, O, O, I**.

i	$w_{i,0}$	**In** $w_{i,1}$	**Out** $w_{i,2}$	**Out** $w_{i,3}$	**In** $w_{i,4}$	**In** $w_{i,5}$	**Out** $w_{i,6}$	**In** $w_{i,7}$
0:	0	4	4	4	0	2	2	6
1:	1	0	6	5	4	0	3	2
2:	2	5	0	6	1	6	0	7
3:	3	1	2	7	5	4	1	3
4:	4	6	5	0	2	3	6	4
5:	5	2	7	1	6	1	7	0
6:	6	7	1	2	3	7	4	5
7:	7	3	3	3	7	5	5	1
$\rho^{t+1}(1)$:	100	010	001	100	010	001	100	010
$w_{i,0}$:	000	100	100	100	000	010	010	110
l:					110	110	110	

Corollary 5.9 is a special case of a more general result that allows random accessing to any window. This more general result will be needed for sequential accessing.

Corollary 5.11. (M-V-W) *Assume a deck of $N = 2^n$ cards starts in initial order and has been perfectly shuffled t times. Then after another k perfect shuffles,*

$$w_{i,t+k} = w_{i,t} \oplus \rho^t(i) \oplus \rho^{t+k}(i) \oplus \bigoplus_{i=1}^{k} \delta(\mathbf{S}_i)\rho^{t+1}(1).$$

Sequential Accessing in a Perfect-Shuffle Memory of Size $N = 2^n$

The breakthrough in extending the accessing algorithms of the perfect-shuffle memory from memories of size $2^n - 1$ to 2^n is due to Stone [**94**]. He observed that certain sequences of out- and in-shuffles cause each data element to appear once and only once in a particular memory location or window. Such a sequence of shuffles is called a *tour*. By mapping the external addresses $0, \ldots, N - 1$ to an internal tour, we can achieve sequential accessing in unit time (after an initial setup).

Consider a memory of size $N = 2^3 = 8$ in initial order. The sequence of shuffles **O-I-O-O-I-I-I** is a tour at windows 3 and 4.

		O	**I**	**O**	**O**	**I**	**I**	**I**
0:	0	0	2	2	2	0	1	5
1:	1	4	0	3	6	2	0	1
2:	2	1	6	0	3	4	3	4
3:	**3**	**5**	**4**	**1**	**7**	**6**	**2**	**0**
4:	**4**	**2**	**3**	**6**	**0**	**1**	**5**	**7**
5:	5	6	1	7	4	3	4	3
6:	6	3	7	4	1	5	7	6
7:	7	7	5	5	5	7	6	2
time:	0	1	2	3	4	5	6	7

FIGURE 6.   Tours through windows (memory locations) 3 and 4.

The external-internal map associates an external address with the pair, $(w_{k,t}, t(\bmod n))$, where $w_{k,t}$ comes from a tour. For the memory of size $8 = 2^3$ shown in figure 6 and the tour passing through window 4, we would use the following correspondences.

external:	0	1	2	3	4	5	6	7
$w_{4,t}$:	4	2	3	6	0	1	5	7
t:	0	1	2	3	4	5	6	7
$t(\bmod 3)$:	0	1	2	0	1	2	0	1

In this example, external address 0 is mapped to $(4, 0)$, 1 to $(2,1)$, 2 to $(3,2)$, and so on.

To sequential access in unit time in a memory of size 2^n, we must find a tour through some window k and associate external addresses with the pairs $(w_{k,t}, t(\bmod n))$. Assume the time is t_0 and we want to access a block of data starting with external address l associated with $(w_{k,t}, t_l)$. First perform $t_l - t_0 (\bmod n)$ shuffles (out or in); the time is now congruent to $t_l (\bmod n)$. Now perform the n-long sequence of shuffles from Corollary 5.11 that random accesses l to window k. Datum l is now at window k at time $t_l (\bmod n)$. Next we "pick up the tour" through window k, performing those shuffles that access $l + 1, l + 2$, and so on to window k. Getting the memory to time $t_l (\bmod n)$ requires an average of $(n - 1)/2$ shuffles, random accessing requires $n = \log_2 N$ shuffles, and sequential accessing occurs in unit time after the first element in a block is accessed. Thus for a perfect-shuffle memory of size 2^n, a block of size q can be accessed in $q + 3(n - 1)/2$.

Consider a memory of size $2^3 = 8$, with read/write port at window 4, $t_0 = 0$, and $w_{4,0} = 0$. Assume we want to access the data block [4, 5, 6, 7]. External address 4 is associated with internal address 0 at time 1(mod 3), 5 is associated with (1, 2), 6 with (5, 0), and 7 with (7, 1). Given these associations, we want to access internal data [0, 1, 5, 7]. We first perform $t_l - t_0 = 1 - 0$ out-shuffles to get the time to 1(mod 3). (We could have performed in-shuffles just as well.) Then we access internal datum 0 to window 4. Finally we pick up the tour and access the rest of the block. This is illustrated in figure 7.

	O	I	O	I	I	I	I
0:	4	4	6	6	2	0	1	5
1:	5	0	4	7	6	2	0	1
2:	6	5	2	4	3	4	3	4
3:	7	1	0	5	7	6	2	0
4:	0	6	7	2	**0**	**1**	**5**	7
5:	1	2	5	3	4	3	4	3
6:	2	7	3	0	1	5	7	6
7:	3	3	1	1	5	7	6	2
$t(\bmod n)$:	0	1	2	0	1	2	0	1

	$t_l - t_0$ out-shuffles	Access 0 to window 4	Pick up the tour to bring 1, 5, 7 to the r/w port

FIGURE 7. The accessing algorithm for a perfect-shuffle memory of size 2^3.

Properties of Tours

The accessing algorithm for a block of data in a perfect-shuffle memory takes about $(\log_2 N)/2$ more than the theoretical minimum ($\log_2 N$ plus 1 for each item in the block)—a nice accomplishment. The success of the algorithm, however, depends upon the existence of a tour for a given memory size. Therein lies a problem: we don't know how to find tours except by exhausting over all **O-I** sequences at a window. There are necessary conditions for the existence of a tour in a window, and these help reduce the number of windows that need to be checked, but the question of the existence of tours is still open.

Theorem 5.12. (M-V-W) *If* $N = 2^n - 1$, *then* $\mathbf{S}_1, \ldots, \mathbf{S}_{N-1}$ *is a tour at window* k *iff* $\mathbf{S}_1', \ldots, \mathbf{S}_{N-1}'$ *is a tour at window* $N - 1 - k$, *where* $\mathbf{O}' = \mathbf{I}$ *and* $\mathbf{I}' = \mathbf{O}$.

In odd decks, the "inverse" of a tour at k is a tour at $N - 1 - k$.

Theorem 5.13. (M-V-W) *If* $N = 2^n$, *then* $\mathbf{S}_1, \ldots, \mathbf{S}_{N-1}$ *is a tour at window* k *iff it is a tour at window* $N - 1 - k$.

In even decks, a tour at k is a tour at $N - 1 - k$.

Corollary 5.14. (M-V-W) *For all* N, *there are the same number of tours at windows* k *and* $N - 1 - k$.

Theorem 5.15. (M-V-W) *If* $N = 2n, \mathbf{I}, \ldots, \mathbf{I}$ *is a tour at every window iff* $2^N \equiv 1 (\bmod N + 1)$ (figure 8).

		I	I	I	I	I	I	I	I	I
0:	0	5	2	6	8	9	4	7	3	1
1:	1	0	5	2	6	8	9	4	7	3
2:	2	6	8	9	4	7	3	1	0	5
3:	3	1	0	5	2	6	8	9	4	7
4:	4	7	3	1	0	5	2	6	8	9
5:	5	2	6	8	9	4	7	3	1	0
6:	6	8	9	4	7	3	1	0	5	2
7:	7	3	1	0	5	2	6	8	9	4
8:	8	9	4	7	3	1	0	5	2	6
9:	9	4	7	3	1	0	5	2	6	8

FIGURE 8. The sequence $I, \ldots, I$ is a tour through every window for $N = 10$.

Definition. If p is an n-digit binary integer, $b_n(p) = p_{n-1}, \ldots, p_1, p_0$, then rev$(p)$ is the integer with reversed binary digits, that is, $b_n(\text{rev}(p)) = p_0, p_1, \ldots, p_{n-1}$. For example, rev$(13) = 22$; $b_5(13) = (0, 1, 1, 0, 1)$; $b_5(\text{rev}(13)) = (1, 0, 1, 1, 0)$.

Theorem 5.16. (M-V-W) *If $N = 2^n$, then $\mathbf{S}_1, \ldots, \mathbf{S}_{N-1}$ is a tour at window k iff it is a tour at window* rev(k).

In decks of size 2^n, the reverse of a tour at k is a tour at rev(k). This information lets us greatly reduce the number of windows we need to check for tours. We apply these results to a deck of size $2^4 = 16$.

k	$b_4(k)$	$N - 1 - k$	$b_4(\text{rev}(k))$	rev(k)
0	0000	15	0000	0
1	0001	14	1000	8
2	0010	13	0100	4
3	0011	12	1100	12
4	0100	11	0010	2
5	0101	10	1010	10
6	0110	9	0110	6
7	0111	8	1110	14
8	1000	7	0001	1
9	1001	6	1001	9
10	1010	5	0101	5
11	1011	4	1101	13
12	1100	3	0011	3
13	1101	2	1011	11
14	1110	1	0111	7
15	1111	0	1111	15

Consider window 1. If there's a tour at 1, then there's one at $15 - 1 = 14$ and one at $8 = \text{rev}(1)$. And if there's a tour at 8, then there's one at $7 = 15 - 8$. Thus if we check window 1 for tours, we don't need to check windows 7, 8, or 14. Using this information on the existence of tours in a deck of size $2^4 = 16$, we reduce the set of windows to check to 0, 1, 2, 3, 5, 6, and 7.

Our final result on the existence of tours comes from examining the binary representations of the tour elements passing through a window. In figure 9 we let $w_{k,t}(i)$ be the ith bit, the 2^i's digit, $0 \le i \le n - 1$ in the binary representation

		O	I	O	O	I	I	I
0:	0	0	2	2	2	0	1	5
1:	1	4	0	3	6	2	0	1
2:	2	1	6	0	3	4	3	4
3:	3	5	4	1	7	6	2	0
4:	**4**	**2**	**3**	**6**	**0**	**1**	**5**	**7**
5:	5	6	1	7	4	3	4	3
6:	6	3	7	4	1	5	7	6
7:	7	7	5	5	5	7	6	2
time:	0	1	2	3	4	5	6	7
$w_{4,t}(2)$	1	**0**	**0**	1	0	0	1	**1**
$w_{4,t}(1)$	0	1	**1**	**1**	0	0	0	1
$w_{4,t}(0)$	0	0	1	**0**	**0**	1	1	1

FIGURE 9.   Binary representations of the tour through window 4.

of $w_{k,t}$. For window 4 in particular, note the block structure of the kth level of bits for each i.

The blocks (and partial blocks) for window 4 are either $(0, 0, 1)$ or $(1, 1, 0)$. If we had followed the tour through window 3, the blocks would be $(0, 1, 1)$ or $(1, 0, 0)$.

Theorem 5.17. (M-V-W) *For a perfect-shuffle memory of size* $N = 2^n$, *let* $\mathbf{S}_1, \ldots, \mathbf{S}_{N-1}$ *be a sequence of shuffles applied to the initial order. Then for each* i, $0 \le i \le n - 1$, *the sequence* $\{w_{k,t}(i)\}$ *of the* ith *bit of the datum in window* k *at time* t *has the following structure:*

(a) *The first* $n - i$ *bits are* $k_i, \ldots, k_{n-1}$;

(b) *These are followed by* s *successive blocks of length* n, *each block having one of two forms* $(k_0, \ldots, k_{n-1})$ *or* $(k_0 \oplus 1, \ldots, k_{n-1} \oplus 1)$;

(c) *The final block is of length* m, $(k_0, \ldots, k_{m-1})$ *or* $(k_0 \oplus 1, \ldots, k_{m-1} \oplus 1)$.

The values s *and* m *satisfy* $N - (n - i) = sn + m$, *with* $0 \le m \le n - 1$.

This theorem provides the foundation of the last result we have about the existence of tours. If there is a tour through window k, every value $0, \ldots, 2^n - 1$ appears exactly once, and there must be 2^{n-1} 0s and 2^{n-1} 1s in the sequences $\{w_{k,t}(i)\}$. Our strategy is to eliminate those windows k where $(k_0, \ldots, k_{n-1})$

or $(k_0 \oplus 1, \ldots, k_{n-1} \oplus 1)$ cannot be combined as blocks to produce half 0s and 1s.

Definition. If p is an n-digit binary integer, $b_n(p) = p_{n-1}, \ldots, p_1, p_0$, then the *density of p is* $d(p) = \sum_{i=1}^{n-l} p_i$, the *density of the beginning $n - l$ bits of p is* $d_{B,l}(p) = \sum_{i=l}^{n-1} p_i$, and the *density of the ending l bits of p is* $d_{E,l}(p) = \sum_{i=0}^{l-1} p_i$.

As an example, consider 13 as a 4-digit binary number: $b_4(13) = (1, 1, 0, 1)$; $d(13) = 3$, $d_{B,2} = 2$, and $d_{E,2} = 1$.

Theorem 5.18. (M-V-W) *In a memory of size 2^n, if a tour exists at window k, then for every $l, 0 \le l \le n - 1$, there exists an integer x such that one of the following equations hold:*

$$(2d(k) - n)x = 2^{n-1} - [s(n - d(k)) + d_{B,l}(k) + d_{E,m}(k)]$$

or

$$(2d(k) - n)x = 2^{n-1} - [s(n - d(k)) + d_{B,l}(k) + m - d_{E,m}(k)],$$

where $2^n - (n - l) = sn + m, 0 \le m \le n - 1$.

As an example of Theorem 5.18, let $N = 2^5 = 32, k = 16$, and $i = 0$, so that

(a) $b_5(p) = b_5(16) = (k_4, k_3, k_2, k_1, k_0) = (1, 0, 0, 0, 0)$;
(b) $d(k) = k_4 + k_3 + k_2 + k_1 + k_0 = 1$;
(c) $N - (n - i) = 32 - (5 - 0) = 27 = s \cdot 5 + m \Rightarrow s = 5$ and $m = 2$;
(d) $d_{B,l}(k) = \sum_{i=l}^{n-1} k_i = 1$; and
(e) $d_{E,l}(k) = \sum_{i=0}^{l-1} k_i = 0$.

With this information in hand concerning the density of the block structure of $\{w_{p,t}(i)\}$, we can see if it's possible to have $\{w_{p,t}(i)\}$ have half 0's and half 1's. We calculate

$$2^{n-1} - [s(n - d(k)) + d_{B,l}(k) + d_{E,m}(k)]$$
$$= 16 - [5(5 - 1) + 1 + 0] = -5,$$
$$2_{n-1} - [s(n - d(k)) + d_{B,l}(k) + m - d_{E,m}(k)]$$
$$= 16 - [5(5 - 1) + 1 + 2 - 0] = -7.$$

Theorem 5.18 says that if a tour exists at window 16, then there is an integer x such that

$$-3x = -5 \quad \text{or} \quad -3x = -7,$$

both of which are impossible. Hence, there can be no tour at window 16 in a memory of size 32.

These combined techniques can significantly reduce the windows that must be searched. As another example, consider a memory of size $N = 2^6 = 64$. Theorem 5.13 eliminates the need to check windows 32–63, and Theorem 5.15 eliminates window 0. Using the conditions of Theorem 5.18, we find that tours may exist only at windows 3, 12, 15, 21, 22, 25, and 26. Theorem 5.16 tells us that a tour at window 3 is a tour at rev(3) = 48, and by Theorem 5.13 a tour at window 48 is a tour at window $64 - 48 - 1 = 15$. We need only check windows 3 or 15, and in a like manner we need only check windows 22 or 26. Checking windows 3, 12, 21, 22, and 25 determines the complete tour structure for a memory of size 64. Only one tour exists, and this is at window 22. The exact tour is listed in Appendix 4.

By similar arguments to those used for $N = 64$, the complete tour structure for $N = 128$ can be determined by examining these nineteen windows: 7, 11, 13, 14, 19, 21, 22, 25, 26, 28, 29, 30, 37, 38, 41, 42, 46, 49, and 54. Since $\mathbf{I}^2$ fixes the datum at window 42, this seems like a good window to check, because no tour can contain the subsequence -$\mathbf{II}$-. An exhaustive search on window 42 found no tours. The issue of tours at the other windows for $N = 128$ is still open.

A few open questions remain about tours.

- Are there other necessary conditions for the existence of $\mathbf{I}$–$\mathbf{O}$ tours?

- What are the necessary and sufficient conditions for the existence of $\mathbf{I}$–$\mathbf{O}$ tours on arbitrary-sized memories, and in particular, memories of size 2^n?

- A (rather tedious) method is given in [78] for extending a tour from a memory of size $16 = 2^4$ to one of size $256 = 2^8$. Can this method be generalized to construct tours on memories of size 2^n for arbitrary n?

- Are there other sets of permutations that allow random accessing in $\log_2 N$ and sequential accessing in unit time? If so, is their tour structure easily determined?

Epilogue

As explained in the introduction of this chapter, the study of dynamic memories faded away when fast, cheap disks and low-power memory with battery back-up were developed to provide nonvolatile storage. The perfect-shuffle memory remains an elegant solution to a problem overtaken by events. Shuffles, however, are still important as computer interconnection networks. A parallel computer with N processors interconnected by ranks of perfect shuffles can move data from one processor to another in $\log_2 N$ steps. The Omega-network, developed by Duncan H. Lawrie, is an example of such a shuffle interconnection.

Trick 5.19. ("Unshuffled," by Paul Gertner) (Note: The previous tricks have been tied back into their chapter by the underlying mathematics. This trick has no mathematical connection to Chapter 5—it's just plain good. It is the most sophisticated application I know of the result $o(\mathbf{O}, 52) = 8$.)

The magician has a card chosen—the 8 of hearts—and returned to the deck. He shows some illegible "edge markings" on the side of the deck. After a faro-shuffle, the edge markings repeat "UNSHUFFLED" four times. After a second faro-shuffle, the word appears twice, and a third faro-shuffle produces the word once in large letters. The deck is then shown to be indeed "unshuffled"—in its original order from the factory. When the spectator reveals the chosen card is the eight of hearts, the magician turns the deck around and shows that the other side reads "EIGHT OF HEARTS."

Explanation

This may be the cleverest trick exploiting the fact that eight out-shuffles returns a deck of size 52 to its original order. The seed of the idea for this trick was planted by Ted Anneman in *The Jinx*, no. 19 (April 1936). He explained how secret messages were sent during World War I written on the side of a deck of cards. "Only the person knowing the order could put them together to make the message readable" [3]. Anneman proposed using this idea to maintain a secret list that could be easily hidden and restored. Thirty-seven years later in an article in *Genii*, Michael Ewer developed this idea into a trick which revealed a spectator's name on the side of a deck through repeated faro shuffles [26]. Nine issues of *Genii* later Bob Wicks proposed several variations of Ewer's trick with different words to be restored on the side of the deck, including *unshuffled*

[**101**]. All these ideas were brilliantly brought together by Paul Gertner in his trick "Unshuffled." Earlier, more complicated versions of "Unshuffled" were published in Gertner's lecture notes and his 1982 *Best of Friends I* [**33**]. The final evolution of "Unshuffled" was published in Richard Kaufman's 1994 book of Gertner's tricks, *Steel and Silver* [**45**]. This simplified version is used here with Gertner's permission. Readers interested in a more sophisticated (and difficult) presentation should consult *Steel and Silver*.

The trick begins by preparing the deck. Set the cards in new deck order, from front to back: Spades (Ace to King), diamonds (Ace to King), Clubs (King to Ace), and Hearts (King to Ace). Wrap the deck tightly with several rubber bands and square the cards precisely. Carefully print the word UNSHUFFLED on one side of the deck with an indelible black marker. On the other side write EIGHT OF HEARTS in similar block letters. Take the 2 of spades from the deck and trim the upper left and lower right corners to make a "corner short card" (see figure 3.1) and return it to its position. Give the deck five out-shuffles and return it to its case. You're ready to start. (In the explanation, the magician's dialogue or "patter" is shown in **_bold italic type_**.)

FIGURE 10.   The unreadable eight copies of "UNSHUFFLED" at the start.

People always ask me if I use marked cards for my tricks. I only do one trick with marked cards, because a marked deck can be pretty suspicious. This deck is "edge marked," which is a little harder to notice than other systems.

During this, remove the deck from the case, being careful to hold the cards so your hand conceals the long side of the deck on which "EIGHT OF HEARTS" is written. After the preliminary five shuffles, "UNSHUFFLED" appears eight times (figure 10) and is unreadable unless you know what you're looking for.

I'm going to have you choose a card—red, black, face or number—it doesn't matter. As I riffle the deck, say "Stop" sometime to get your card.

After the five out-shuffles, the deck doesn't appear to be in any order, and the corner-shortened two of spades is now right above the 8 of hearts. Hold the deck in dealing position in the left hand and riffle the outer-left corner with the left thumb. When you gently riffle down the cards from top to bottom, you'll find the cards naturally stop with the two of spades above your thumb and the eight of hearts on the top of the bottom portion of the deck. Practice the timing, and it will look like you genuinely stopped when the spectator asked. Cut the cards above your left thumb to the table. Remove the top card from the packet in your left hand and show it to the audience.

Here's the card you stopped at; remember its name.

Show the eight of hearts and replace it. Pick up the cards on the table and put them on top of those in your left hand. The deck is now restored to its original order—five out-shuffles from new deck order and three out-shuffles before that order is restored.

Your card will be mixed in the deck with what I call an "Un" shuffle. It's a strange concept—like negative entropy—but once you understand the theory, the possibilities are amazing.

Do an out-shuffle and square the deck.

To the casual observer, it looks like the entropy is increasing—the cards appear to be shuffled, but the reality is just the opposite. We can verify this by looking at the edge markings on the side of the deck. They're starting to read "UNSHUFFLED" four times.

Be sure to hold the cards so the word is clearly shown to the audience, but so they can't see the words "EIGHT OF HEARTS" on the back (figure 11).

We can further decrease the entropy by unshuffling again. It looks like a very precise shuffle, but because every card in the pack is controlled, the natural randomness evaporates. Here, the edge markings show the progress—it reads "UNSHUFFLED" twice.

Do another out-shuffle.

FIGURE 11.   Showing "UNSHUFFLED" four times.

We conclude this little experiment with one final unshuffle. The edge markings give us the status—the deck is now completely unshuffled. That's what it says on the side. However, every well-designed experiment should allow for independent verification. If the deck says "UNSHUFFLED," every

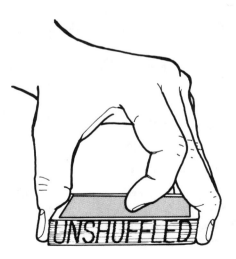

FIGURE 12.   The deck after three out-shuffles, showing "UNSHUFFLED."

card should be in perfect sequence, Ace through King, hearts, clubs, spades, and diamonds.

Perform a third out-shuffle. Square the cards and show the side that reads "UNSHUFFLED," as in figure 12. Be careful that no one sees the words "EIGHT OF HEARTS" on the other side. Turn the pack face up, spread the cards, and show they're in perfect order.

This has been a good experiment in theoretical physics, but it's not a good magic trick—yet. I still have to find the card you chose at the beginning of the trick.

Loosely square up the cards and turn the deck face down with the "EIGHT OF HEARTS" side towards the spectators. The deck should be unaligned, so the words can't be read on the side (figure 13). This is in preparation of the climax.

FIGURE 13. The unsquared deck with illegible wording being slowly squeezed together.

What was the name of your card?

The answer is, "Eight of Hearts." Now a little theatrics will produce a grand climax. Pretend you've mis-heard the name of the card. With a grand flourish turn over the top card and say,

Here it is, right on top, the Ace of Hearts.

The spectator should let you know you goofed and that the card was the *eight* of hearts. Give a puzzled look and put the Ace of Hearts on top of the unsquared deck.

FIGURE 14.   The climax of "Unshuffled," the chosen card revealed.

Hmmm... this could be a problem. But if we really need an eight of hearts, we can get it without much more work.

Take the unsquared deck with the "EIGHT OF HEARTS" side facing the audience. Slowly square the deck. It will appear that the word "UNSHUF-FLED" is transformed into "EIGHT OF HEARTS," as in figure 14. After the gasps subside, brace yourself for a round of wild applause. Well, I'm sure that's the way it always happens with me!

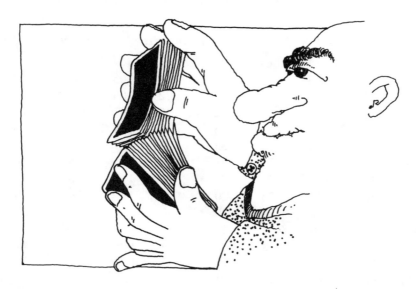

Appendix 1
The Order of Shuffles

Listed below are the orders of Shuffles, both in- and out-, for any deck of size 3–200. (The orders for decks of size 1 and 2 are left as an exercise for the reader.) If N is odd, $o(\mathbf{O}, N) = o(\mathbf{I}, N)$, and $2^{o(\mathbf{O},N)} \equiv 1 \pmod{N}$. If N is even, $o(\mathbf{O}, N) = o(\mathbf{O}, N-1)$, and $o(\mathbf{I}, N) = o(\mathbf{O}, N+1)$. There is no closed formula for $o(\mathbf{O}, N)$, but $o(\mathbf{O}, 2n)|\lambda(2n-1)|\varphi(2n-1)$, where λ is Carmichael's function, and φ is Euler's totient function.

N	$o(\mathbf{O}, N)$	$o(\mathbf{I}, N)$	N	$o(\mathbf{O}, N)$	$o(\mathbf{I}, N)$	N	$o(\mathbf{O}, N)$	$o(\mathbf{I}, N)$
3	2	2	12	10	12	21	6	6
4	2	4	13	12	12	22	6	11
5	4	4	14	12	4	23	11	11
6	4	3	15	4	4	24	11	20
7	3	3	16	4	8	25	20	20
8	3	6	17	8	8	26	20	18
9	6	6	18	8	18	27	18	18
10	6	10	19	18	18	28	18	28
11	10	10	20	18	6	29	28	28

N	$o(\mathbf{O},N)$	$o(\mathbf{I},N)$	N	$o(\mathbf{O},N)$	$o(\mathbf{I},N)$	N	$o(\mathbf{O},N)$	$o(\mathbf{I},N)$
30	28	5	67	66	66	104	51	12
31	5	5	68	66	22	105	12	12
32	5	10	69	22	22	106	12	106
33	10	10	70	22	35	107	106	106
34	10	12	71	35	35	108	106	36
35	12	12	72	35	9	109	36	36
36	12	36	73	9	9	110	36	36
37	36	36	74	9	20	111	36	36
38	36	12	75	20	20	112	36	28
39	12	12	76	20	30	113	28	28
40	12	20	77	30	30	114	28	44
41	20	20	78	30	39	115	44	44
42	20	14	79	39	39	116	44	12
43	14	14	80	39	54	117	12	12
44	14	12	81	54	54	118	12	24
45	12	12	82	54	82	119	24	24
46	12	23	83	82	82	120	24	110
47	23	23	84	82	8	121	110	110
48	23	21	85	8	8	122	110	20
49	21	21	86	8	28	123	20	20
50	21	8	87	28	28	124	20	100
51	8	8	88	28	11	125	100	100
52	8	52	89	11	11	126	100	7
53	52	52	90	11	12	127	7	7
54	52	20	91	12	12	128	7	14
55	20	20	92	12	10	129	141	4
56	20	18	93	10	10	130	14	130
57	18	18	94	10	36	131	130	130
58	18	58	95	36	36	132	130	18
59	58	58	96	36	48	133	18	18
60	58	60	97	48	48	134	18	36
61	60	60	98	48	30	135	36	36
62	60	6	99	30	30	136	36	68
63	6	6	100	30	100	137	68	68
64	6	12	101	100	100	138	68	138
65	12	12	102	100	51	139	138	138
66	12	66	103	51	51	140	138	46

N	o(**O**, N)	o(**I**, N)	N	o(**O**, N)	o(**I**, N)	N	o(**O**, N)	o(**I**, N)
141	46	46	161	33	33	181	180	180
142	46	60	162	33	162	182	180	60
143	60	60	163	162	162	183	60	60
144	60	28	164	162	20	184	60	36
145	28	28	165	20	20	185	36	36
146	28	42	166	20	83	186	36	40
147	42	42	167	83	83	187	40	40
148	42	148	168	83	156	188	40	18
149	148	148	169	156	156	189	18	18
150	148	15	170	156	18	190	18	95
151	15	15	171	18	18	191	95	95
152	15	24	172	18	172	192	95	96
153	24	24	173	172	172	193	96	96
154	24	20	174	172	60	194	96	12
155	20	20	175	60	60	195	12	12
156	20	52	176	60	58	196	12	196
157	52	52	177	58	58	197	196	196
158	52	52	178	58	178	198	196	99
159	52	52	179	178	178	199	99	99
160	52	33	180	178	180	200	99	66

Appendix 2
How to Do the Faro Shuffle

The actual performance of the faro shuffle—even using an old and battered deck—is many times easier than verbal or written instructions. Nonetheless, instructions and illustrations are here for O_2—the double faro—for a right-handed performer. The method shown works for me, but it certainly isn't the only way to hold the cards. Several other methods are illustrated at the end of the description.

The type of cards used may be as important as anything else.

- Start practicing with bridge-sized cards, which are about 0.25-inch narrower than poker-sized cards. The smaller cards are easier to learn with, especially if you have small hands.
- Do not use plastic or plastic-coated cards; they are not flexible enough.
- Do not use "cute" bridge cards; the paper stock is not suitable.
- Do use almost any type of U.S. Playing Card Company card, such as Aviator, Bicycle, or Bee.

The Double or Ordinary Faro Shuffle

Holding the Deck

Hold the deck in the left hand, face of the deck to the palm, little finger supporting, thumb on one side, ring, middle, and index fingers opposite. Sometimes during the shuffle it may help to curl the index finger against the face of the packet in the left hand to keep the cards together. Grasp the deck with the thumb and middle finger of the right hand, with the right index finger curled against the back of the deck (figure 1).

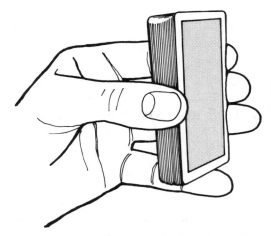

FIGURE 1.   Holding the deck for doing a faro shuffle.

Cutting the Deck

By careful estimation, cut the deck in half, keeping the grip unchanged with the left hand (figure 2). Cutting the deck exactly in half is perhaps the most difficult part of faro shuffling. An even deck is easier to cut in half than an odd deck, because an error in an even deck produces a difference of at least two cards in the halves. In a deck of fifty-two, for example, instead of 26–26 you might get 25–27 or 24–28. This thickness of two cards is enough for most people to discern, while a thickness of one card is more difficult.

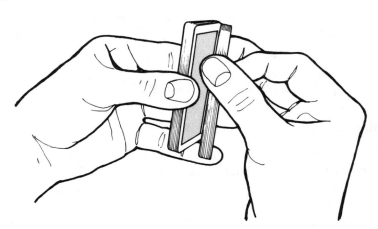

FIGURE 2.   Cutting off half the cards.

Gripping the Halves

Place the right index finger on top of the deck, with the thumb on one side and ring, middle, and little fingers opposite. The left little and right index fingers support the bottom and top of the opposite halves and provide needed tension. Turn the two packets at right angles and square them against each other (figure 3).

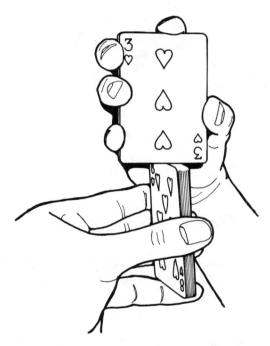

FIGURE 3. Squaring the halves against each other. Note placement of fingers of right hand.

Starting the Shuffle

Place the inner corners of the packets together, gripping them firmly, and keeping them square with the left little and right index fingers. An in- or out-shuffle is started at this point, depending on where the top card of the right

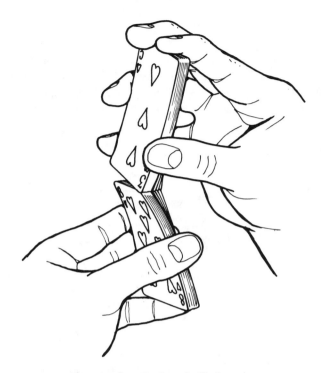

FIGURE 4. The faro shuffle in action.

hand half is placed. Apply light pressure causing the two halves to bend slightly (figure 4).

Finishing the Shuffle

Carefully rocking the cards and gently releasing the pressure causes them to spring together interlaced. Continued pressure and rocking will interlace the cards from back to front. Sometimes a card will be slightly out of line or bent or sticky, which may cause it to not go into place. If so, back the halves apart to the recalcitrant card and very carefully reapply pressure, perhaps guiding the cards together with a thumb.

The importance of releasing the cards was recognized over a century ago when John Maskelyne described the process (figure 5).

> In faro, the manner of dealing the cards necessarily divides them into two equal parts. This being the case, they are taken up by the dealer, one in each

hand. Holding them by the ends, he presses the two halves together so as to bend them somewhat after the manner shown ..., in the position "A." The halves are now "wriggled" from side to side in opposite directions, with what would be called in mechanism a "laterally reciprocating motion." This causes the cards to fly up one by one, from either side alternately, as indicated in the figure at "B" [**67**].

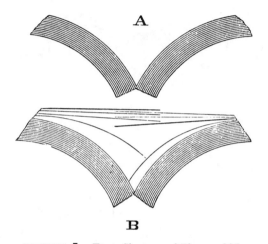

FIGURE 5. From *Sharps and Flats*, p. 205.

Here's another description from Michael Close, a contemporary magician, on the all-important release of pressure.

The weaving of the cards is not the result of pressure; **it is the result of the release of pressure**, accompanied with a small inward twist of the two packets. This is the biggest mistake that people make, because they try to force the two packets together. The inner corners of the two halves are brought together, and the right hand exerts a small pressure which is directed in and [down]. The corners of the packets will bend very slightly due to this pressure. Then the pressure is released, and at the same time the right inner corner twists outward. The cards will begin to weave, and as the pressure is released from [back] to [front] all the cards will follow suit. This is the "knack" part of the shuffle, and it is the reason why you will be struggling with the shuffle, and then all of a sudden it will work. What you've done is release the pressure [**14**].

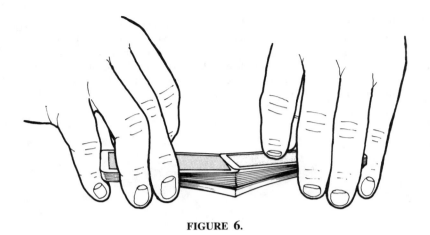

FIGURE 6.

Other Shuffle Methods

Figure 6 illustrates the method recommended by *The Greater Magic Library* for the perfect shuffle [**37**]. Ed Mishell, a renowned magic illustrator, suggested the technique pictured in Figure 7, with the little fingers providing support to the two halves [**70**].

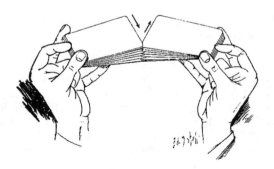

FIGURE 7.

FIGURE 8.

Harry Lorayne recommended doing the shuffle from an entirely different angle. He suggested using the right index finger to control the pressure on the cards (figure 8) [**56**].

Jean Hugard recommended getting support for the deck halves from a table rather than from fingers (figure 9) [**40**].

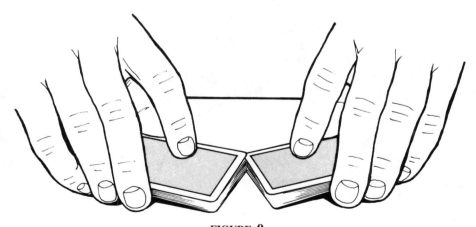

FIGURE 9.

THE TRIPLE FARO SHUFFLE

I developed this method of performing a triple faro shuffle in 1973, while I was working on my dissertation research at Duke University. It served more as the solution to a challenge (could I do a triple faro), than as a significant research aid. As far as I know, no one else has ever published a comparable method or even realized it could be performed. Karl Fulves, a prolific student of the shuffle, said "[The Triple faro] represents something of a problem in that the shuffle itself is impossible to perform with ordinary techniques [27]." Of course, after you see what's required, it may be clear why no one else bothered to invent it. The description that follows assumes a 51-card deck.

Cutting Off the Top Third

I could never master "eye-balling" one-third of the deck as I can with one-half, so I just count off seventeen cards. Hold the deck face down in the left hand in a dealing position with the left thumb resting on the outer left corner. Place the right index finger on the back of the deck to help hold it together. Slightly bevel the outer left corner and riffle off seventeen cards (figure 10).

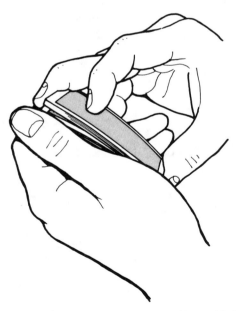

FIGURE 10.   Counting off seventeen cards by riffling with the left thumb.

The First Shuffle

In-shuffle the seventeen cards into the remaining thirty-four and push them in about 1 inch (figure 11).

FIGURE 11. In-shuffling the top third into the bottom two-thirds.

Cutting Off the Bottom Third

Put your left index finger on the face of the bottom card of the protruding top third; push off all cards below it. There are exactly seventeen cards to be pushed off (figure 12).

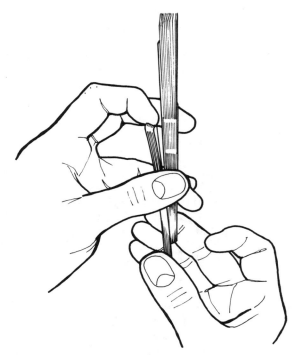

FIGURE 12.   Cutting off the bottom third of the deck by pushing off all cards below the face of the top third.

The Second Shuffle

Tightly grip the shuffled-together top and middle thirds of the deck. Squeeze the cards so that the bottoms of the cards in the middle third are held together. Square the bottom and middle thirds against each other (figure 13), and in-shuffle the bottom of the middle third into the bottom third of the deck (figure 14). (The top of the middle third is shuffled into the top third.) At this point you are ready to do the triple incomplete control for trick 4.6, "The Triple Seekers."

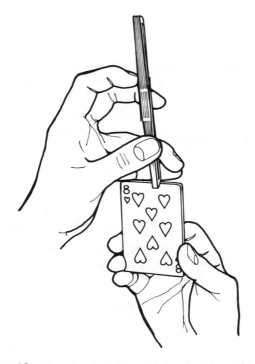

FIGURE 13.   Squaring the bottom third against the middle third.

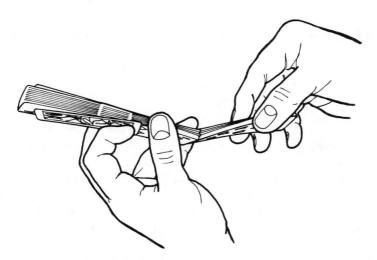

FIGURE 14.   Squaring the bottom third against the middle third.

The Rotation

What follows is not for the faint of heart! Rotate the top third at right angles to the middle third of the deck (figure 15). Put the left middle finger in the angle between the top and middle thirds. The middle finger curls around to grip the back of the middle third, pushing inwards on the cards. The left index finger curls against the face of the top third, pushing outwards. The pressure between the index and middle fingers tips keeps the top third in position.

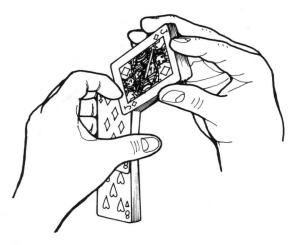

FIGURE 15.   Rotating the top third of the partially shuffled deck at right angles to the rest.

The Third Shuffle

Pivot the bottom third up, keeping it shuffled into the middle third. Continue to hold the top third at right angles to the middle third with the pressure between the left index and middle fingers. Slowly move the corner of the bottom third towards the edge of the top third. (All the while the middle third is still shuffled into both top and bottom thirds.) The object is to in-shuffle the top third into the bottom third. Slowly push the top and bottom thirds together (figure 16). Gently twist or bend the corner of the bottom third to help push the into the gaps between the cards of the top third (figure 17). This is the most tedious part

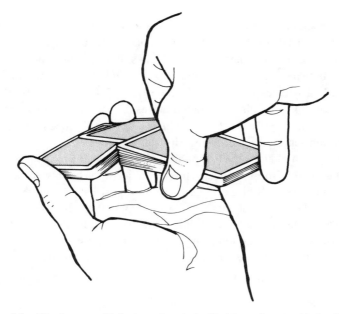

FIGURE 16. The bottom third pivoted and shuffled into the top third, while both remain shuffled into the middle third.

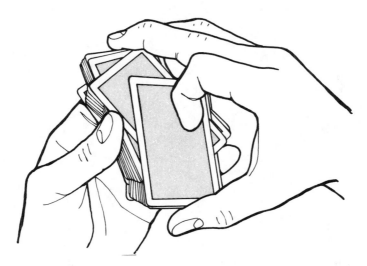

FIGURE 17. The three parts of the deck being pushed together after a triple shuffle.

of the triple shuffle, and I don't know how to do it (or describe it) any better. Sometimes I must do the final shuffle almost one card at a time. However, when you're finished, you've joined the limited ranks of triple shufflers.

Assuming the deck starts in standard order, $[0, 1, 2, \ldots, 49, 50]$, the shuffle just described will leave the deck in order $[34, 17, 0, 35, 18, 1, 36, 19, 2, \ldots, 49, 33, 15, 50, 34, 16]$. This permutation moves the card in position p to $3p + 2$ (mod 52). By varying whether the top third is in- or out-shuffled into the middle third, you can achieve each of the 3! triple faro shuffles.

Murray Bonfeld took the first published step down the road to performing a triple shuffle. He almost duplicated this technique when he solved the problem of putting the top cards of a deck at every third position. He did the first two shuffles described here, but didn't follow through with the final shuffle. Rather he just pushed the top and bottom thirds together and let them interlace in a chance manner. "Place the long side of the cards against the table and jiggle the cards together" [9]. The top third alone was spaced correctly.

Appendix 3
Tours on Decks of Size 8, 16, 32, and 64

A tour is a sequence of shuffles that causes each card in a deck to appear once and only once in a particular location or window. For example, in a memory of size $N = 8$, the shuffle sequence **OIIIOII** brings each card to the windows 1 and 6.

		O	**I**	**I**	**I**	**O**	**I**	**I**
0:	0	0	2	3	7	7	6	2
1:	**1**	**4**	**0**	**2**	**3**	**5**	**7**	**6**
2:	2	1	6	1	6	3	4	3
3:	3	5	4	0	2	1	5	7
4:	4	2	3	7	5	6	2	0
5:	5	6	1	6	1	4	3	4
6:	**6**	**3**	**7**	**5**	**4**	**2**	**0**	**1**
7:	7	7	5	4	0	0	1	5

TABLE 1 $N = 8$

Tour		Window	
OIOOI	**II**	3	4
OIIIO	**OI**	1	6
OIIIO	**II**	1	6
IOOII	**IO**	3	4
IIOII	**IO**	3	4
IIIOO	**IO**	1	6

TABLE 2 $N = 16$

Tour			Window			
00010	01001	00011	3	6	9	12
00010	01100	01001	3	6	9	12
00010	01101	01011	3	12		
00011	00100	01100	3	6	9	12
00011	01011	00011	3	12		
00011	10001	00100	3	6	9	12
00011	10001	10101	3	12		
00011	10101	01100	3	12		
00100	01100	10001	3	6	9	12
00100	01110	00100	3	6	9	12
00100	01110	10101	3	12		
00100	10001	11000	3	6	9	12
00100	10010	00111	3	6	9	12
00100	11000	10011	3	6	9	12
00100	11010	10111	3	12		
00110	00100	11000	3	6	9	12
00110	01000	11001	3	6	9	12
00110	10101	11000	3	12		
00110	10110	00111	3	12		
00111	00010	01001	3	6	9	12
00111	00011	01011	3	12		
00111	01010	11001	3	12		
01000	11001	00011	3	6	9	12
01000	11100	01001	3	6	9	12
01000	11101	01011	5	12		
01001	00011	10001	3	6	9	12
01001	00100	01110	3	6	9	12
01001	10001	00110	3	6	9	12
01001	10101	01110	3	12		
01010	11001	00011	3	12		
01010	11100	01001	3	12		
01010	11101	01011	3	5	10	12
01011	00011	10001	3	12		

TABLE **2**   (*continued*)

Tour			Window			
O I O I I	O O I O O	O I I I O	3	12		
O I O I I	I O O O I	O O I I O	3	12		
O I O I I	I O I O I	O I I I O	3	5	10	12
O I O I I	I I I O I	O I I O I	1	14		
O I O I I	I I I O I	O I I I I	1	7	8	14
O I I O O	O I O O I	I O O O I	3	6	9	12
O I I O O	O I I I O	O O I I O	3	12		
O I I O O	I O O O I	I O O I O	3	6	9	12
O I I O O	I O O O I	I O O I O	3	12		
O I I O I	O I O I I	I O O O I	3	12		
O I I O I	O I I O O	O I I I O	3	12		
O I I I O	O O I O O	I O O I O	3	6	9	12
O I I I O	O O I O O	I I O I O	3	12		
O I I I O	O O I I O	I O I I O	3	12		
O I I I O	I O I O I	I O O I O	3	12		
O I I I O	I O I O I	I I O I O	3	5	10	12
O I I I I	I O I O I	I I I I O	1	7	8	14
I O O O I	O O I O O	I O O O I	3	6	9	12
I O O O I	O O I I O	O O I O O	3	6	9	12
I O O O I	O O I I O	I O I O I	3	12		
I O O O I	I O O I O	O O I I O	3	6	9	12
I O O O I	I O I O I	I O O O I	3	12		
I O O O I	I I O O O	I O O I O	3	6	9	12
I O O O I	I I O O O	I I O I O	3	12		
I O O O I	I I O I O	I O I I O	3	12		
I O O I O	O O I I O	O I O O O	3	6	9	12
I O O I O	O O I I I	O O O I O	3	6	9	12
I O O I O	O O I I I	O I O I O	3	12		
I O O I O	O I O O O	I I I O O	3	6	9	12
I O O I O	I I I I I	O I O I I	7	8		
I O O I I	O O O I O	O I I O O	3	6	9	12
I O O I I	O I O I O	I I I O O	3	12		
I O I O I	O I I O O	I O O O I	3	12		

TABLE 2   (*continued*)

Tour			Window			
IOIOI	OIIIO	OOIOO	3	12		
IOIOI	OIIIO	IOIOI	3	5	10	12
IOIOI	OIIII	IOIOI	7	8		
IOIOI	IOOOI	IIOOO	3	12		
IOIOI	IOOIO	OOIII	3	12		
IOIOI	IIOOO	IOOII	3	12		
IOIOI	IIOIO	IOIII	3	5	10	12
IOIOI	IIIIO	IOIOI	1	14		
IOIOI	IIIIO	IOIII	1	7	8	14
IOIIO	OOIII	OOOII	3	12		
IOIIO	OIOOO	IIIOI	3	12		
IOIIO	IOIII	IIOIO	7	8		
IOIII	OOOIO	OIIOI	3	12		
IOIII	OIOIO	IIIOI	3	5	10	12
IOIII	OIOII	IIIOI	7	8		
IOIII	IIOIO	IIIOI	1	14		
IOIII	IIOIO	IIIII	1	7	8	14
IIOOO	IOOIO	OIOOO	3	6	12	
IIOOO	IOOII	OOOIO	3	6	9	12
IIOOO	IOOII	OIOIO	3	12		
IIOOO	IIOIO	IIOOO	3	12		
IIOOO	IIIOO	OIIOI	3	12		
IIOOI	OOOII	OOIOO	3	6	9	12
IIOOI	OOOII	IOIOI	3	12		
IIOIO	IOIIO	OIOOO	3	12		
IIOIO	IOIII	OOOIO	3	12		
IIOIO	IOIII	OIOIO	3	5	10	12
IIOIO	IIOOO	IIIOO	3	12		
IIOIO	IIIII	OIOOI	1	14		
IIOIO	IIIII	OIOII	1	7	8	14
IIIOO	OIOOI	OOIOO	3	6	9	12
IIIOO	OIOOI	IOIOI	3	12		
IIIOO	OIIOI	OIIOO	3	12		

TABLE 2　(*continued*)

Tour			Window			
I I I O I	O I O I I	O O I O O	3	12		
I I I O I	O I O I I	I O I O I	3	5	10	12
I I I O I	O I I I I	I O I O I	1	7	8	14
I I I O I	O I I I I	I O I I I	1	14		
I I I O I	I I I I O	I O I I I	7	8		
I I I I O	I O I I I	I I O I O	1	7	8	14
I I I I I	O I O I I	I I I O I	1	7	8	14
I I I I I	O I O I I	I I I I I	1	14		
I I I I I	I I O I O	I I I I I	7	8		

TABLE 3　$N = 32$

Tour							Window	
I O I O I	I O O I I	O I O I I	O O I I O	I O O I O	I I O I I	I	11	20
I O I O I	I I I I O	I O O I O	I I I I I	O I O I I	O I I I I	I	11	20
I I I O I	I O I O O	I O I I O	O I I O I	O I I O O	I I O I O	I	5	26
I I I I I	O I I O I	O I I I I	I O I O O	I O I I I	I I O I O	I	5	26

TABLE 4　$N = 64$

Tour									Window	
I O O O O	I O O O I	I I O O O	I O I I I	O O I I O	I I O I O					
O I O I O	I I O O I	O I O O I	I I I O O	I I O I O	O I O I O	I I I			22	41
I I I O I	O I O O I	O I I O O	I I I I O	O I O I O	O I I O I					
O I O O I	O I I O I	I O O I I	I O I O O	O I I I O	O O I O O	O O I			26	37

Appendix 4
A Lagniappe

This appendix should be an unexpected bonus, like the extra joker in a deck of cards. Three tricks are explained here, none requiring a faro shuffle or sleight of hand. This does not mean, however, that you should do them without some practice! The first trick uses ternary notation and the special design of this book. The second uses the inverse shuffle on sixteen cards. The third uses another design feature of the book and a special function, $f(x) = 14$.

The Book of Theorems

Alex Elmsley, the early pioneer of faro shuffle mathematics, invented many clever tricks. One of his commercial effects was marketed in 1973 as *The Book of Fortunes* by Ken Brooke's Magic Place; it is explained in [**69**]. *The Book of Theorems* is my modification (and perhaps refinement) of Elmsley's invention. I was helped by comments from two colleagues, Jack C. Mortick and F. Ben Cole. This trick employs some simple mathematics and a cleverly designed book (this one) to yield a startling effect.

The Effect. The magician brings out two decks of cards—one for himself and one for the spectator. Before starting the trick, the magician's deck is "tuned" to the spectator's personality. This is accomplished by the spectator answering three questions; after each answer, the magician drops a few cards on the table. Finally the spectator's lucky number is found by both the magician and spectator dealing cards from their decks until there's a match. Turning to the lucky number page in this book, the spectator finds an image of the matching cards and his answers to all three questions.

Explanation. Both decks are pre-arranged or stacked with eighteen cards on top, and the answers to the questions determine how many cards are dropped on the table. To understand how the trick works, we consider the trick with just eight cards. Assume we have stacked the same eight cards on top of each deck and the magician's deck has been out-shuffled. If zero cards are taken from the top of the magician's deck, then both decks match in position zero. If one card is taken, then they match in position one, and so on. The situations for removing zero to seven cards from the magician's deck are shown in the diagram below. Boldface numbers indicate where the decks match. The spectator's and magician's decks are designated by S and M.

Removed from M deck:	0	1	2	3	4	5	6	7
	0 0	0 –	0 1	0 –	0 2	0 –	0 3	0 –
	1 –	**1 1**	1 –	1 2	1 –	1 3	1 –	1 4
	2 1	2 –	**2 2**	2 –	2 3	2 –	2 4	2 –
	3 –	3 2	3 –	**3 3**	3 –	3 4	3 –	3 5
	4 2	4 –	4 3	4 –	**4 4**	4 –	4 5	4 –
	5 –	5 3	5 –	5 4	5 –	**5 5**	5 –	5 6
	6 3	6 –	6 4	6 –	6 5	6 –	**6 6**	6 –
	7 –	7 4	7 –	7 5	7 –	7 6	7 –	**7 7**
	– 4	– –	– 5	– –	– 6	– –	– 7	– –
	– –	– 5	– –	– 6	– –	– 7	– –	– –
	– 5	– –	– 6	– –	– 7	– –	– –	– –
	– –	– 6	– –	– 7	– –	– –	– –	– –
	– 6	– –	– 7	– –	– –	– –	– –	– –
	– –	– 7	– –	– –	– –	– –	– –	– –
	– 7	– –	– –	– –	– –	– –	– –	– –
	S M	S M	S M	S M	S M	S M	S M	S M

The trick actually uses eighteen cards rather than eight as in the example above. Put these eighteen cards on the top of each deck:

2♣-7♥-A♠-8♦-2♥-8♠-A♦-8♣-2♦-8♥-4♠-5♦-9♣-10♥-4♦-5♣-2♠-3♦

Now give the magician's deck an out-shuffle. (If you can't do a faro shuffle, just put one card between 2♣ and 7♥, 7♥ and A♠, A♠ and 8♦, and so on.) You're set up and ready for the trick.

Bring out both decks of cards and this book. Casually give the spectator his deck and set the book aside. Take your deck out of its case.

I'm going to show you a strange relationship between math, cards, and fortunes. The math is in this book on **Magic Tricks, Card Shuffling, and Dynamic Computer Memories,** *and the cards will help me tell your fortune. First I must tune the deck to your thoughts. You'll have to answer a few simple questions. Here's the first one.*

If [something] happens, will you be happy, sad, or indifferent?

As you follow the instructions below, you're told to drop some cards on the table; these are always taken from the top of the deck. Hold the deck in your left hand for dealing and casually push off the required number of cards with your left thumb. Take the counted off cards with your right hand and drop them in a pile on the table (Southpaws, invert the instructions).

This first question is variable, and can be adapted to any current event like the Super Bowl or the World Series or an election.

- If they will be happy: drop zero cards on the table.
- If they will be sad: drop one card on the table.
- If they will be indifferent: drop two cards on the table.

Here's the second question. What's your favorite fruit: apple, banana, or grapes?

- If they say apple: drop zero cards on the table.
- If they say banana: drop three cards on the table.
- If they say grapes: drop six cards on the table.

Now here's the last question, even simpler than the first two. Which card suits do you like better: red (hearts and diamonds) or black (spades and clubs)?

- If they say red: drop zero cards on the table.
- If they say black: drop nine cards on the table.

Recapitulation. Before the trick, both decks were stacked with the cards shown above; the magician's deck was given one out-shuffle. The spectator's deck and this book have been set aside on the table. The magician takes out his deck and "tunes" it by asking three questions and dropping for the first question, 0, 1, or 2 cards; for the second question 0, 3, or 6 cards, and for the last question, 0 or 9 cards. Thus a total of n cards, $0 \le n \le 17$, cards have been dropped on the table. The two decks now have matching cards at position n (or $n + 1$ if you start counting with origin 1).

The icons and cards printed on pages 1–18 of this book correspond to the answers to the three questions and the card stack just given. They are summarized in the table following. The icons and cards on the other pages are purely decorative.

Cards Dropped	3rd Answer	2nd Answer	1st Answer	Card
0	♦♥	apple	:)	card
1	♦♥	apple	:(	card
2	♦♥	apple	:|	card
3	♦♥	banana	:)	card
4	♦♥	banana	:(	card
5	♦♥	banana	:|	card
6	♦♥	grapes	:)	card
7	♦♥	grapes	:(	card
8	♦♥	grapes	:|	card
9	♣♠	apple	:)	card
10	♣♠	apple	:(	card
11	♣♠	apple	:|	card
12	♣♠	banana	:)	card
13	♣♠	banana	:(	card
14	♣♠	banana	:|	card
15	♣♠	grapes	:)	card
16	♣♠	grapes	:(	card
17	♣♠	grapes	:|	card

The last step in the trick is to find your lucky number. Take your deck out of its case and we'll both deal cards until we find a match. I think you'd agree that's pretty lucky (though its probability of happening is about $1 - 1/e$, which is pretty high for just luck).

Pick up the cards you've dropped on the table and put them on the bottom of your deck. The only thing required of the magician now is a little showmanship. With n cards dropped off the deck in response to the three questions, there will be matching cards at the "lucky" number $n + 1$ (count the first card as one, not zero). On that page in this book is printed a miniature of the card plus icons to answer the three questions.

Well, this is interesting. Your lucky number is n and this must be your lucky card. Let's check page n and see what the book has to say about it.

The Mathematician, the Psychic, and the Magician

I am indebted to Jon Racherbaumer for the very clever handling of this trick with its triple climax. Its origins are not clear, but the basic principle was shown in 1986 when Paul Swinford published "The Discerning Joker" in [**96**]. Swinford used a special deck of cards (available from Haines' House of Cards in Cincinnati) that were binary encoded with holes punched in their edges. He used a knitting needle to quickly sort the cards for the tricks.

The Effect. The magician shows sixteen cards. The spectator thinks of a number and then counts down to that card. The magician shows eight of the cards and asks if the selected card is among them. This is repeated three more times. The chosen card is then shown to be on top of the packet. The spectator counts down to the position to find a second card. This card has been predicted by the magician, and it is the only different-backed card in the packet.

Explanation. This is a clever, self-working trick that uses Theorem 5.10. (The actual use of the theorem is left as an exercise for the reader.) Take fifteen cards from a deck and add a sixteenth, different card from a different deck. Put the different-backed card on top of the sixteen card packet, turn the cards face up, but be careful the spectators don't see its back. Write the name of the different-backed card on a piece of paper and seal it in an envelope. You're ready to go!

Have you ever noticed how different professions approach problems differently? A mathematician and a computer scientist and an engineer can each solve a problem, but they'll approach it three different ways. I'm going to show you a trick with sixteen cards and how a mathematician, a psychic, and a magician would handle it. Let's start with a mathematician's approach to the trick.

Spread out the cards on the table so each one can be seen. Casually drop the sealed envelope off to the side, but make no comment about it. It will be used later.

Think of a number between one and sixteen; remember it. Mentally count to the card at that position and remember the card. You've got two things to remember—a number and a card—don't forget them. Now I'm going to ask you four yes–no questions to find your card.

Be sure to indicate that the counting starts with the different-backed card, which is now the bottom card of the spread-out face-up packet. Pick up the cards and spread them in a face-up fan. Partially pull out of the fan every other card starting with the top card (figure 1).

FIGURE 1.   An inverse faro shuffle with every-other card pulled out.

Is your card among these eight I've pulled out?

After the spectator answers, completely pull out the eight cards. If the chosen card is among the eight you pulled out, put them underneath the other cards. Otherwise put them on top. Repeat this three more times. Spread out the cards face up when you're finished. What you've done is a radix sort of the sixteen cards which requires $\log_2 16 = 4$ inverse shuffles. These inverse shuffles move the chosen card from position n to the top, and the top card (with a different back) to position n (figure 2). The hard work is now over; all that's left is the showmanship.

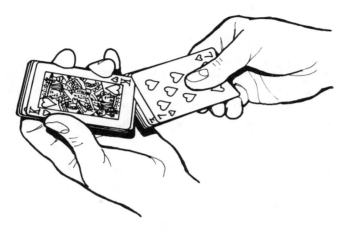

FIGURE 2. Pulling out the the cards from an inverse faro shuffle.

A mathematician would point out that the card you selected is now the top card of the sixteen. This is clever, even surprising, but it smacks of algorithmic manipulation of the cards. It just seems like the mathematician knows some sneaky formula that you don't.

Now that's not the way a psychic would approach the trick. A psychic predicts the future. Here's a sealed envelope that's been on the table from the start, and it contains my psychic impression of what will happen next. What was the number you thought of? Let's count down to that position and see what card is there.

The different-backed card will be in this position. Push it out of the spread and casually turn over the other cards.

Open the envelope. You'll see that the card you chose was predicted in advance, which is exactly the way a psychic does business. A magician, however, goes beyond what a mathematician or a psychic can do. A magician doesn't just find a card or predict the future, he does the impossible—like changing the color of the back of your card.

As you start talking about what a magician does, pick up the different-backed card. Give it a snap and turn it over as you explain the impossible, as if you just changed the color of the card. This triple climax should leave your audience stunned.

A Constant Function

Magicians use all sorts of specially prepared or "gaffed" items for their tricks. It's essential that the audience isn't aware of the gaff, if the trick is to succeed. A particularly clever gaff is a forcing book where the nth word of each page or chapter is the same. In this book, for example, the fourteenth word of the preface, the introduction, the acknowledgments, the bibliography, each chapter, and each appendix is *deck*. To create a minor miracle you just need a function with $f(x) = 14$ for all x.

I want to try a little experiment in predicting the future. I'll put my prediction over here for later reference. Take this book, a brilliant example of fine technical exposition, and pick any of the chapters or appendices.

Write down the word *deck* on a card and put it aside. Let the spectator choose a chapter.

Now let's pick a word at random from that chapter. Roll this pair of dice and add the numbers on the top and bottom. Count to that word, and let's see how it compares to my prediction.

Bibliography

Selected Perfect Shuffle References

If I have seen further than others, it is by standing on the shoulders of giants.
Sir Isaac Newton

This bibliography contains over one hundred references to the perfect shuffle—on a deck of cards or in a computer circuit or as an abstraction. Several references are difficult-to-find magic publications. Many of these had limited distribution when published because few magicians have mastered the perfect shuffle. Some of these books had such small printings that they were mimeographed or photocopied, and certainly cannot be considered publications of record. Under the best of circumstances, libraries carry only magic books geared for general audiences; the specialized books usually are bought and read only by the "inner circle." Nonetheless these books are important in giving a fuller picture of the attempts to understand the perfect shuffle.

My relationship to these references is that of a persistent accumulator; the hard work was done by others. Whenever I find an article on shuffling, I always check its references for ones I don't have. Similarly I always check electronic databases for shuffle references. Having built this bibliography slowly over several decades, I worry that I may not recognize (or remember) everyone who has contributed. Some, however, have made such substantial contributions to the bibliography of shuffling that they must be recognized. Martin Gardner sent me my first dozen or so shuffle references in the early 1970s when I was working on my dissertation. He sent me a few more while I was working on this book. Jack Potter's *The Master Index of Magic in Print* supplied most of the magic references, though I relied on Richard Kaufman's research on the genealogy of Paul Gertner's trick, "Unshuffled." Persi Diaconis did the pioneering work that uncovered most of the early and French references; his shuffle bibliography is given in his joint paper with Ron Graham and William Kantor. Persi also pointed me toward the references on Marlo's incomplete faro control. There's a nice history of the perfect shuffle with several historical references in Jeff Busby's *Epoptica* book review of Jon Racherbaumer's *Card Finesse*.

[1] Adler, Irving. "Make up your own card tricks." *Journal of Recreational Mathematics*, vol. 6, Spring 1973, pp. 87–91.

[2] Aho, A. and J. D. Ullman. "Dynamic memories with rapid random and sequential access." *IEEE Transactions on Computers*, vol. C-20, no. 3, Mar. 1974, pp. 272–276.

[3] Anneman, Ted. *The Jinx*, no. 19, Apr. 1936.

[4] Anonymous. *A Grand Exposé of the Science of Gambling*. New York: Frederic A. Brady, 1860.

[5] —— *Koschitz's Manual of Useful Information*. Kansas City, Mo.: McClintock & Koschitz, Pub., 1894.

[6] —— *The Whole Art and Mystery of Modern Gaming Fully Expos'd and Detected*. London: J. Roberts, 1726.

[7] Bayer, Dave and Persi Diaconis. "Trailing the dovetail shuffle to its lair." *The Annals of Applied Probability*, vol. 2, no. 2, 1992, pp. 294–313.

[8] Bivens, Irl C., ed. "Student research projects." *College Mathematics Journal*, vol. 22, no. 2, Mar. 1991, pp. 144–147.

[9] Bonfield, Murray. "A solution to Elmsley's problem." *Genii*, vol. 37, May 1973, pp. 195–196.

[10] —— "Placements for thirds," in *Faro Concepts*. [N.J.]: [Karl Fulves], 1977, pp. 31–32.

[11] Braun, John. "T. Nelson Downs & Edward 'Tex' McGuire Hocus Pocus Parade." *Linking Ring*, vol. 51, no. 4, Apr. 1971, pp. 53–83; no. 5, May 1971, pp. 61–95.

[12] Busby, Jeff. Review of *Card Finesse* by John Racherbaumer. *Epoptica*, May 1982, pp. 38–44.

[13] Chen, P.Y., D. H. Lawrie, P.-C. Yew, and D. A. Padua. "Interconnection networks using shuffles." *Computer*, Dec. 1981, pp. 55–64.

[14] Close, Michael. *Workers Number 5*. [Carmel, Ind.]: [The Author], 1996.

[15] Davio, Marc. "Kronecker products and shuffle algebra." *IEEE Transactions on Computers*, vol. C-30, no. 2, Feb. 1981, pp. 116–125.

[16] De Moivre, Abraham. *The Doctrine of Chances*. 3rd ed. London: Millar, 1756. Reprinted N.Y.: Chelsea, 1967.

[17] Diaconis, Persi, James Allen Fill, and Jim Pitman. "Analysis of top to random shuffles." *Combinatorics, Probability and Computing*, vol. 1, 1992, pp. 135–155.

[18] Diaconis, Persi, R. L. Graham, and William M. Kantor. "The mathematics of perfect shuffles." *Advances in Applied Mathematics*, vol. 4, 1983, pp. 175–193.

[19] Diaconis, Persi, Michael McGrath, and Jim Pitman. "Riffle shuffles, cycles, and descents." *Combinatorica*, vol. 15, 1995, no. 1, pp. 11–29.

[20] Duck, J. Russell. "Rusduck 'stay-stack' system." *Cardiste*, no. 1, Feb. 1957, p. 12.

[21] Eisemann, Kurt. "Number theoretic analysis and extensions of 'the most complicated and fantastic card trick ever invented." *American Mathematical Monthly*, vol. 91, no. 5, May 1984, pp. 284–289.

[22] Elmsley, Alex. "The mathematics of the weave shuffle." *Pentagram*, vol. 11, Jun. 1957, pp. 70–71, Jul. 1957, pp. 78–79, Aug. 1957, p. 85.

[23] —— "Work in progress." *Ibidem*, no. 11, Sept. 1957, p. 21.

[24] Epstein, Richard A. *The Theory of Gambling and Statistical Logic*. Academic Press, 1967.

[25] Euler, Leonhard. "Sur l'advantage du banquier au jeu de pharaon." *Histoire de l'Academie Royale des Sciences et Belles-Lettres*. vol. for 1764, publication date 1766.

[26] Ewer, Michael S. "A name revelation with faro shuffles." *Genii*, vol. 37, no. 11, Nov. 1973, pp. 465–468.

[27] Fulves, Karl. *Faro Possibilities*. [N.J.]: The author, 1966.

[28] —— *Faro and Riffle Technique*. 3rd printing. [N.J.]: The author, 1976.

[29] Ganson, Lewis. *Dai Vernon's Ultimate Secrets of Card Magic*. Tahoma, Calif.: L&L Publishing, 1995.

[30] Gardner, Martin. "Mathematical games: Can the shuffling of cards (and other apparently random events) be reversed?" *Scientific American*, vol. 215, no. 4, Oct. 1966, pp. 114–117.

[31] Gardner, Martin and C. A. McMahan. "Riffling casino checks." *Mathematics Magazine*, vol. 50, no. 1, Jan. 1977, pp. 38–41.

[32] Gertner, Paul. *Best of Friends, I*. Pittsburgh: The author, 1982.

[33] Golomb, Solomon W. "Permutations by cutting and shuffling." *SIAM Review*, vol. 3, no. 4, Oct. 1961, pp. 293–297.

[34] Green, J. H. *An Exposure of the Arts and Miseries of Gambling*. Cincinnati: U. P. James, 1843.

[35] Herstein, I. N. *Topics in Algebra*. Waltham, Mass.: 1964

[36] Heuer, C. V. Solution to "Lost in the Shuffle," problem E 2318, "Elementary problems and solutions." *American Mathematical Monthly*, vol. 79, no. 8, Oct. 1972, p. 912.

[37] Hilliard, John N. *The Greater Magic Library*. 5 vols. Carl Waring Jones and Jean Hugard, eds. N.Y.: A.S. Barnes and Co., 1956.

[38] "How to win at poker, and other science lessons." *The Economist*, Oct. 12, 1996, pp. 87–88.

[39] Hugard, Jean. "Weaving the cards," in *Card Manipulations No. 3*. 1934, p. 96.

[40] Hugard, Jean, and Frederick Braue. "The perfect faro shuffle," in *Expert Card Technique*. Minneapolis: Carl Waring Jones, 1940, pp. 143–155.

[41] Innis, S. Victor. *Inner Secrets of Crooked Card Players*. Los Angeles: The author, 1915.

[42] Johnson, Paul B. "Congruences and card shuffling." *American Mathematical Monthly*, vol. 63, Dec. 1956, pp. 718–719.

[43] Jordan, Charles T. *Thirty Card Mysteries*. 2nd rev. ed. Penngrove, Calif.: The author, 1919, 1920.

[44] —— "Trailing the dovetail shuffle to its lair." *The Bat*, no. 59, Nov. 1948, p. 429; no. 60, Dec. 1948, p. 440; no. 61, Jan. 1944.

[45] Kaufman, Richard. *Paul Gertner's Steel and Silver*. Silver Spring, Md.: Kaufman and Greenberg, 1994.

[46] Kolata, Gina. "Perfect shuffles and their relation to math." *Science*, vol. 216, 1982, pp. 505–506.

[47] —— "Prestidigitator of digits." *Science 85*, vol. 6, no. 3, Apr. 1985, pp. 66–72.

[48] Lenfant, J. "Fast random and sequential access to dynamic memories of any size." *IEEE Transactions on Computers*, vol. C-25, no. 1, Jan. 1976.

[49] Le Paul, Paul. *The Card Magic of LePaul*. Danville, Ill.: The Interstate, Printers & Publishers, 1949.

[50] Lévy, Paul. "Étude d'une classe de permutations." *Comptes Rendus Académie des Sciences*, vol. 227 (1948), pp. 422–423.

[51] —— "Étude d'une nouvelle classe de permutations." *Comptes Rendus Académie des Sciences*, vol. 227 (1948), pp. 578–579.

[52] —— *Quelques Aspects de la Pensée d'un Mathématicien*. Paris: Libraire Scientifique et Technique, Albert Blanchard, 1970.

[53] —— "Sur deux classes de permutations." *Comptes Rendus Académie des Sciences*, vol. 228 (1949), pp. 1089–1989.

[54] —— "Sur quelque classes de permutations." *Compositio Mathematica*, vol. 8 (1950), pp. 1–48.

[55] —— "Sur un classe remarquable de permutations." *Bulletin Académie Royal de Belgique*, vol. 2 (1949), pp. 361–377.

[56] Lorayne, Harry. "Faro shuffle," in *Close Up Card Magic*, 1962, pp. 16–18.

[57] Mann, Brad. "How many times should you shuffle a deck of cards?" *Journal of Undergraduate Mathematics and Its Applications*, vol. 15, no. 4, Winter 1994, pp. 303–332.

[58] Marlo, Edward. "The centre card transfer." *The Gen*, vol. 16, no. 6, Oct. 1960, pp. 130–131.

[59] —— "The drawkcab [backward] faro." *Ibidem*, no. 12, Dec. 1957, p. 4.

[60] —— "The faro calculator." *Ibidem*, no. 12, Dec. 1957, p. 2.

[61] —— *Faro Controlled Miracles*. Chicago: The author, 1964.

[62] —— "Faro notes," chapter 7 of *Revolutionary Card Techniques*. Chicago: The author, 1958.

[63] —— "The faro shuffle," chapter 6 of *Revolutionary Card Techniques*. Chicago: The author, 1958.

[64] —— "The incomplete faro," *The New Tops*, vol. 3, no. 9, Sept. 1963, pp. 18–19.

[65] —— "No binary today," *The Gen*, vol. 16, no. 6, Oct. 1960, pp. 131–132.

[66] —— "Remember and forget," *The New Tops*, vol. 3, no. 8, Aug. 1963, pp. 26–28.

[67] Maskelyne, John N. *Sharps and Flats*. 2nd ed. London: Longmans, Green & Co., 1894. Reprint. Las Vegas: G.B.C. Press, no date.

[68] Medvedoff, Steve and Kent Morrison. "Groups of perfect shuffles." *Mathematics Magazine*, vol. 60, no. 1, Feb. 1987, pp. 3–15.

[69] Minch, Stephen. *The Collected Works of Alex Elmsley*. 2 vols. Tahoma, Calif.: L & L Publishing, 1991.

[70] Mishell, Ed. "The troublesome weave shuffle." *M-U-M*, vol. 47, no. 9, Feb. 1958, p. 387.

[71] Montmart, Pierre Remond de. *Essai d'Analyse sur les Jeux de Hazards*. Paris: J. Quillan, 1708.

[72] Morris, S. Brent. "The basic mathematics of the faro shuffle." *Pi Mu Epsilon Journal*, vol. 6, no. 2, Spring 1975, pp. 85–92.

[73] —— "Faro shuffling and card placement." *Journal of Recreational Mathematics*, vol. 8, no. 1, Winter 1975, pp. 1–7.

[74] —— and Robert E. Hartwig. "The generalized faro shuffle." *Discrete Mathematics*, vol. 15, no. 4, Aug. 1976, pp. 333–345.

[75] —— "Magic tricks, card shuffling, and dynamic computer memories." 48 minutes. N.Y.: Association for Computing Machinery, 1992. Videocassette.

[76] —— "Permutations by cutting and shuffling: A generalization to Q dimensions." Ph.D. Dissertation, Department of Mathematics, Duke University, Durham, N.C., 1974.

[77] —— "Practitioner's commentary: Card shuffling." *Journal of Undergraduate Mathematics and Its Applications*, vol. 15, no. 4, Winter 1994, pp. 303–338.

[78] ——, Arthur Valliere, III, and Richard A. Wisniewski. "Processes for random and sequential accessing in dynamic memories." *IEEE Transactions on Computers*, vol. C-28, no. 3, Mar. 1979, pp. 225–237.

[79] —— *Method and Apparatus for Random and Sequential Accessing in Dynamic Memories*. U.S. Patent 4,161,036. July 10, 1979.

[80] —— and Robert E. Hartwig. "The universal flip matrix and the generalized faro shuffle." *Pacific Journal of Mathematics*, vol. 58, no. 2, Jun. 1975, pp. 445–455.

[81] Packard, Robert W. and Erik S. Packard. "The order of a perfect k-shuffle." *Fibonacci Quarterly*, vol. 32, no. 2, May 1994, pp. 136–144.

[82] Parlett, David. *The Oxford Guide to Card Games*. N.Y.: Oxford University Press, 1990.

[83] Potter, Jack, comp. *The Master Index to Magic in Print*. 14 vols. Calgary, Alberta: Micky Hades Enterprises, 1967–1975.

[84] —— *The Master Index to Magic in Print, Supplements*. Vols. 1–9. Calgary, Alberta: Micky Hades Enterprises, 1971–1975.

[85] Racherbaumer, Jon. *The Artful Dodges of Eddie Fields*. N.p.: N.p., 1968.

[86] Ramnath, Sarnath and Daniel Scully. "Moving card i to position j with perfect shuffles." *Mathematics Magazine*, vol. 69, no. 5, Dec. 1996, pp. 361–365.

[87] Ravelli [Ronald Wohl]. "The mathematics of the weave shuffle." *Ibidem*, no. 36, Mar. 1979, pp. 2–25.

[88] Ronse, Christian. "A generalization of the perfect shuffle." *Discrete Mathematics*, vol. 47 (1983), pp. 293–306.

[89] Rosenthal, John W. "Card shuffling." *Mathematics Magazine*, vol. 54, no. 2, Mar. 1981, pp. 64–67.

[90] —— "Faro favorites." *Cardiste*, no. 2, May 1957, p. 14; no. 3, Jun. 1957, p. 14, no. 4, Sept. 1957, p. 15; no. 5, Feb. 1958, p. 14.

[91] Scarne, John. *Scarne's New Complete Guide to Gambling*. Rev. ed. N.Y.: Simon & Schuster, 1986.

[92] Stone, Harold S. "Parallel processing with the perfect shuffle." *IEEE Transactions on Computers*, vol. C-20, no. 2, Feb. 1971.

[93] —— "Dynamic memories with enhanced data access." *IEEE Transactions on Computers*, vol. C-21, no. 4, Apr. 1972, pp. 359–366.

[94] —— "Dynamic memories with fast random and sequential access." *IEEE Transactions on Computers*, vol. C-24, no. 12, Dec. 1975, pp. 1167–1174.

[95] Sun, Jianli, Eduard Cerny, and Jan Gecsei. "Fault tolerance in a class of sorting networks." *IEEE Transactions on Computers*, vol. 43, no. 7, Jul. 1994, pp. 827–837.

[96] Swinford, Paul. *The Cyberdeck*. Cincinnati: Haines' House of Cards, 1986.

[97] —— *Faro Fantasy*. Connersville, Ind.: Haley Press, 1968.

[98] —— *More Faro Fantasy*. Connersville, Ind.: Haley Press, 1971.

[99] Thompson, J. G., Jr. *My Best*. N.Y.: Louis Tannen, 1959.

[100] Thorp, E.O. "Nonrandom shuffling with applications to the game of faro." *Journal of the American Statistical Association*, vol. 68, 1973, pp. 812–847.

[101] Weilandt, H. *Finite Permutation Groups*. New York: Academic Press, 1964.

[102] Wicks, Robert. "The handwriting on the deck." *Genii*, vol. 38, no. 8, Aug 1974, pp. 330–331.

[103] Wong, C. K. and D. T. Tang. "Dynamic memories with faster random and sequential access." *IBM Journal of Research and Development*, vol. 21, no. 3, 281–288.

[104] Yoshigahara, Nobuyuki. "Card return numbers." *Journal of Recreational Mathematics*, vol. 5, 1972, pp. 36–38.

Index